S0-AWE-985

Applied Mathematical Sciences | Volume 4

J. K. Percus

Combinatorial Methods

With 58 Illustrations

Springer-Verlag New York · Heidelberg · Berlin
1971

Jerome K. Percus

New York University Courant Institute of Mathematics and Sciences New York, New York

Library of Congress Catalog Card Number 78-152001.
Printed in the United States of America.

ISBN 0-387-90027-6 Springer-Verlag New York • Heidelberg • Berlin
ISBN 3-540-90027-6 Springer-Verlag Berlin • Heidelberg • New York

PREFACE

It is not a large overstatement to claim that mathematics has traditionally arisen from attempts to understand quite concrete events in the physical world. The accelerated sophistication of the mathematical community has perhaps obscured this fact, especially during the present century, with the abstract becoming the hallmark of much of respectable mathematics. As a result of the inaccessibility of such work, practicing scientists have often been compelled to fashion their own mathematical tools, blissfully unaware of their prior existence in far too elegant and far too general form. But the mathematical sophistication of scientists has grown rapidly too, as has the scientific sophistication of many mathematicians, and the real world - suitably defined - is once more serving its traditional role. One of the fields most enriched by this infusion has been that of combinatorics. This book has been written in a way as a tribute to those natural scientists whose breadth of vision has inparted a new vitality to a dormant giant.

The present text arose out of a course in Combinatorial Methods given by the writer at the Courant Institute during 1967-68. Its structure has been determined by an attempt to reach an informed but heterogeneous group of students in mathematics, physics, and chemistry. Its lucidity has been enhanced immeasurably by the need to satisfy a very resolute critic, Professor Ora E. Percus, who is responsible for the original lecture notes as well as for their major modifications. The writer would like to thank Professor James Steadman for the arduous task of proof-reading, establishing consistency of notation, and making a number of revisions to improve clarity.

J. K. Percus
New York
June 30, 1971

TABLE OF CONTENTS

I.   COUNTING AND ENUMERATION ON A SET

(Minimal Geometric Structure)

A.   Introduction

1§.   Set Generating Functions

The basic problem of combinatorial mathematics is that of recognizing and enumerating or at least counting objects of specified character, out of an enormous number of unspecified objects. Most often, this is accomplished, implicitly or explicitly, by attaching an algebraic tag or weight to each desired trait and summing the weights thereby obtained. The desired objects can then be identified at leisure.

Let us first consider such weighted sums or "generating functions" for the complete array of subsets of a given set, without further restriction. Even here, the word complete admits of several possibilities which must be spelled out. We deal then with an n-set $S = \{a,b,c,\ldots,z\}$ of n distinguishable elements. n! ordered sets or permutations can thus be described, e.g., $(a,c,z,b,\ldots)$, and written in various forms $(a,1)$, $(b,4)$, $(c,3)$,$\ldots$ or

$$\begin{pmatrix} a & & b \\ & c & \\ & z & \end{pmatrix} \quad \text{or} \quad \begin{matrix} .a \\ .c \\ . \quad .z \\ .b \end{matrix} \quad ,\ldots$$

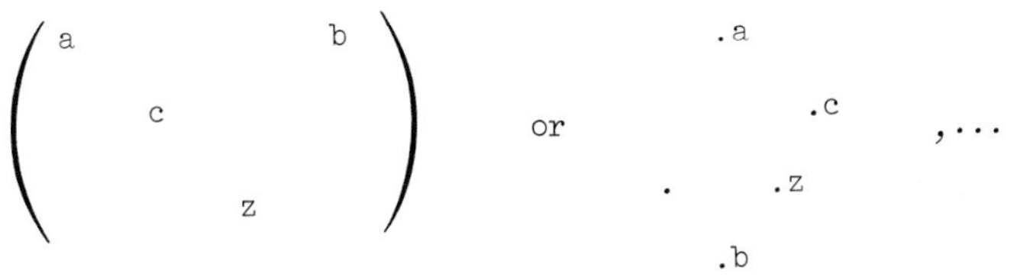

On the other hand, S can be regarded as unordered so that all permutations are to be identified. Now

(i) For S an unordered n-set, the formal subset generator which enumerates all possible subsets of S, is clearly

$$(1+a)(1+b)(1+c) \cdots = 1 + a + b + c + \cdots + ab + ac + \cdots ,$$

where 1 denotes the empty set.

(ii) For S an unordered n-set, the generator for "subsets with repetition" ["subsets with repetition" means sets with elements from S where repetition is allowed, and there are not necessarily subsets of S, e.g., for $S = \{a,b\}$ a subset with repetition weight be $\{a,a,a,b\}$] can following (i) be written as

$$(1+a+aa+\cdots)(1+b+bb+\cdots)\ \cdots = [(1-a)(1-b)\ \cdots]^{-1}.$$

For ordered subsets, there are of course many more possibilities. Now

(iii) The ordered set generator is given by the permanent

$$\mathrm{Per}\begin{pmatrix} a & a & \cdots \\ b & b & \\ c & c & \\ \vdots & \vdots & \cdots \end{pmatrix}$$

where $\mathrm{Per}\,(a_{ij})_{i,j=1,\ldots,n}$ is the expansion of $\mathrm{Det}\,(a_{ij})_{i,j=1,\ldots,n}$ with $(-1)^P$ replaced by $+1$, i.e., we define

$$\mathrm{Per}\,(a_{ij})_{i,j=1..n} \equiv \sum_P \prod_{i=1}^{n} a_{P(i),i}$$

where P denotes a permutation of $(1,\ldots,n)$. Further properties of the permanent will be important in the sequel. At this stage, we observe that since the permanent is linear in each row vector and column vector, it follows that

$$\mathrm{Per}\,(a_{ij}+b_{ij})_{i,j=1\ldots n} = \sum_{s,t} \mathrm{Per}\,(a_{ij})_{i\in s,j\in t}\mathrm{Per}\,(b_{ij})_{i\in\bar{s},j\in\bar{t}}$$

where s,t are subsets of $(1,\ldots,n)$ and $\bar{s},\bar{t}$ the complements of s and t; here s and t are required to have the same order. An easy consequence, together with (iii), is

(iv) The ordered subset generator is given by

$$\mathrm{Per}\left[I + \begin{pmatrix} a & a & \cdots \\ b & b & \cdots \\ \vdots & \vdots & \end{pmatrix}\right] \quad \text{where } I = \begin{pmatrix} 1 & & & O \\ & 1 & & \\ & & \ddots & \\ O & & & 1 \end{pmatrix}.$$

The algebraic difficulties which we will encounter in using the permanent stem directly from its restriction to linear terms in each variable. Thus, matters simplify dramatically when repetition is allowed. To start with,

(v) The generator for ordered m (not necessarily $m = n$) sets is constructed by choosing first any of $(a,\ldots,z)$, then any of $(a,\ldots,z)$, then again any of $(a,\ldots,z)$, i.e., we have

$$(a+b+\cdots+z)(a+b+\cdots+z)\ \cdots\ (a+b+\cdots+z) = (a+b+\cdots+z)^m.$$

(vi) Consequently the generator for all "subsets with repetition" is

$$(a+b+\cdots+z)^0 + (a+b+\cdots+z)^1 + (a+b+\cdots+z)^2 + \cdots = [1 - (a+b+\cdots+z)]^{-1}.$$

2§. Numerical Generating Functions

Given a set generating function, it is in principle possible to find weighted sums over the more stringently specified subsets. These weighted sums are of several types:

(a) a true average occurring in some physical situation

(b) a weight of unity for sets satisfying given conditions, 0 otherwise; hence the number of sets of specified type is counted.

(c) an intermediate situation in which the sets satisfying desired conditions are given a common weight which can be extracted when the remaining sets are weighted in some other prescribed fashion; we then have a numerical generating function.

Examples. Let us consider several trivial and one easy non-trivial example.

How many k-subsets of an n-set exist? To solve that, we weight each element multiplicatively by x so that a k-subset is weighted by x^k. The total weight of generating function according to (i) is $(1+x)^n$. Hence coef x^k in $(1+x)^n = \binom{n}{k}$.

How many k-subsets with repetition exist? Here we look at the generating function of subsets with repetition, i.e., the coefficient of x^k in $(1-x)^{-n}$ (see (ii)) which equals $\binom{n-1+k}{k}$.

How many ordered k-subsets with repetition exist? This is the coefficient of x^k in $(1-nx)^{-1}$ (see (vi)), i.e., n^k.

Fibonacci numbers.

How many non-confluent subsets (adjacent elements not both occupied) are there of n ordered elements? Let $f(n)$ be the number of non-confluent subsets of an n-set. We find a recursion relation for $f(n)$ by considering the state of the last element. If unoccupied, then the previous $n - 1$ elements can be any non-confluent (n-1)-set, and if occupied the (n-1)-th must be unoccupied while the remaining $n-2$ elements can be any non-confluent (n-2)-set. Schematically

$$f(n) \quad = \quad f(n-1) \quad + \quad f(n-2)$$

where 1 indicates that the cell is occupied (i.e., the subset contains the element in question) and 0 indicates that the cell is not occupied. In order to solve the above difference equation we require the obvious boundary conditions,

$$f(0) = 1, \qquad f(1) = 2.$$

Let us carry out the solution by means of the generating function $F(x) = \sum_{j=0}^{\infty} f(j)x^j$. Since

$$x\,F(x) = \sum_{j=0}^{\infty} f(j)\, x^{j+1} = \sum_{j=1}^{\infty} f(j-1)\, x^j$$

and

$$x^2 F(x) = \sum_{j=0}^{\infty} f(j)\, x^{j+2} = \sum_{j=2}^{\infty} f(j-2)\, x^j,$$

then the recursion relation and boundary condition imply:

$$F(x) - 1 - 2x = x\,F(x) - x + x^2 F(x)$$

or

$$F(x) = \frac{1+x}{1-x-x^2} = \frac{\frac{1}{2} + \frac{3}{2\sqrt{5}}}{1 + \frac{x}{\frac{1}{2} - \frac{1}{2}\sqrt{5}}} + \frac{\frac{1}{2} - \frac{3}{2\sqrt{5}}}{1 + \frac{x}{\frac{1}{2} + \frac{1}{2}\sqrt{5}}}\,.$$

Therefore,

$$f(n) = \left(\frac{1}{2} + \frac{3}{2\sqrt{5}}\right)\left(\frac{1}{2} + \frac{\sqrt{5}}{2}\right)^n + \left(\frac{1}{2} - \frac{3}{2\sqrt{5}}\right)\left(\frac{1}{2} - \frac{\sqrt{5}}{2}\right)^n.$$

One of the major uses of generating functions is in the determination of asymptotic forms. E.g., $f(x)$ above has a simple pole of smallest magnitude at $x_0 = -\frac{1}{2} + \frac{1}{2}\sqrt{5}$. Therefore, for x near x_0, $F(x)$ diverges as a geometric series i.e., $\frac{1}{x-x_0} \sim \Sigma\left(\frac{x}{x_0}\right)^n$. We conclude that for large n, $f(n) \sim x_0^{-n} = ((-1+\sqrt{5})/2)^{-n}$ $= \left(\frac{1+\sqrt{5}}{2}\right)^n$ which of course agrees with the exact answer.

More generally, if $F(x) = \sum_{j=0}^{\infty} f(j)x^j$ has a unique simple pole of smallest magnitude at x_0, then

$$f(j) = \frac{1}{2\pi i} \oint_{\bar{c}} \frac{F(x)}{x^{j+1}}\,dx = \frac{1}{2\pi i} \oint_{c} \frac{F(x)}{x^{j+1}}\,dx - \frac{\text{Res } F(x_0)}{x_0^{j+1}} ,$$

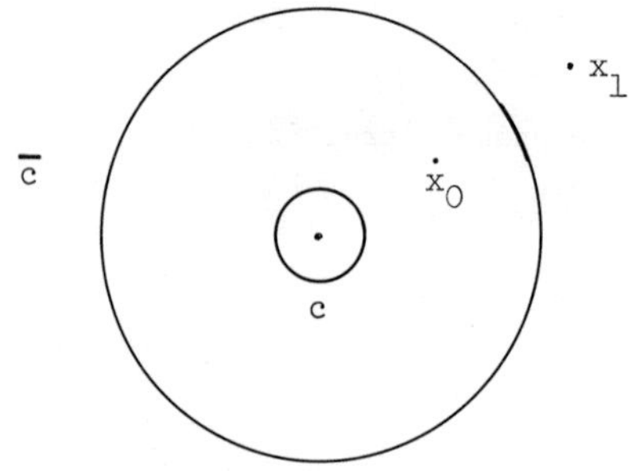

where c is a small circle around the origin, $\bar{c}$ is a circle of radius $\bar{x}$ where $|x_0| < |\bar{x}| < |x_1|$. Hence

$$\left|f(j) + \frac{\text{Res } F(x_0)}{x^{j+1}}\right| = \left|\frac{1}{2\pi i} \oint_{\bar{c}} \frac{F(x)}{x^{j+1}}\,dx\right| \leq$$

$$\leq \frac{1}{2\pi} \oint_{\bar{c}} \left|\frac{F(x)}{x^{j+1}}\right| dx \leq \frac{\max\limits_{x\in\bar{c}}|F(x)|}{|\bar{x}^j|}$$

gives the asymptotic form

$$f(j) = \frac{-\text{Res } F(x_0)}{x_0^{j+1}} (1+k(j))$$

where

$$|k(j)| \leq \frac{x_0 \max\limits_{x\in\bar{c}}|F(x)|}{\text{Res } F(x_0)} \left(\frac{x_0}{\bar{x}}\right)^j .$$

B. Counting with Restrictions -- Techniques

1§. Inclusion-Exclusion Principle

It is almost always easier to count with multiple prohibitions by going over to the complementary permitted situations. This is the inclusion-exclusion relation: if there are N objects, and exactly $N(\alpha)$ of these have property α, $N(\beta)$ have property $\beta, \ldots, N(\alpha,\beta)$ have properties α and β etc. (in each case with the remaining properties undetermined), then the number devoid of all properties is

$$\begin{aligned} N(\alpha',\beta',\gamma',\ldots) = N &- (N(\alpha) + N(\beta) + N(\gamma) + \cdots) \\ &+ (N(\alpha,\beta) + N(\alpha,\gamma) + \cdots) \\ &- (N(\alpha,\beta,\gamma) + \cdots) \\ &+ \cdots \quad . \end{aligned}$$

This follows from $N(\alpha') = N - N(\alpha)$

$$\begin{aligned} N(\alpha',\beta') &= N(\beta') - N(\alpha,\beta') \\ &= N - N(\beta) - (N(\alpha) - N(\alpha,\beta)), \text{ etc.} \end{aligned}$$

More generally if each element x is weighted by $\omega(x)$, and if

$$W(j) = \sum_{x,C} \omega(x|\alpha_1 \ldots \alpha_j)$$

where the summation is over all elements x and all combinations C of j conditions, then the total weight of elements satisfying no conditions is given by

$$E(0) = W(0) - W(1) + W(2) - \ldots \quad .$$

Furthermore, the total weight of elements satisfying exactly k properties of the total of m imposed is:

$$E(k) = W(k) - \binom{k+1}{k}W(k+1) + \binom{k+2}{k}W(k+2) - \cdots + (-1)^{m-k}\binom{m}{k}W(m).$$

For a direct proof: suppose that x of weight $\omega(x)$ satisfies exactly t properties; then in the L.H.S. of the above equation x contributes $\omega(x)$ if $t = k$ and 0 otherwise. Now consider the R.H.S.: If $t < k$, obviously x contributes 0; if $t = k$, clearly $\omega(x)$; for $t > k$, x occurs in $\binom{t}{p}$ sets with p required properties, therefore contributing

$$\begin{aligned}\omega(x)[\binom{t}{k} - \binom{k+1}{k}\binom{t}{k+1} + \binom{k+2}{k}\binom{t}{k+2} - \cdots] \\ = \omega(x)\binom{t}{k}[1 - \binom{t-k}{1} + \binom{t-k}{2} - \cdots] \\ = \omega(x)\binom{t}{k}(1-1)^{t-k} = 0\end{aligned}$$

which completes the proof.

Note too that

$$1 - \binom{t-k}{1} + \binom{t-k}{2} - \cdots + (-1)^p\binom{t-k}{p} = (-1)^p\binom{t-k-1}{p},$$

so that the remainder when stopping at the p-th term of $E(k)$ always has the sign of the next term: each term always overcompensates.

Note that the expression for $E(k)$ leads to a simple relation between generating functions

$$\begin{aligned}\sum_k E(k)t^k &= \sum_{k,\ell} (-1)^\ell \binom{k+\ell}{k} W(k+\ell)t^k \\ &= \sum_{k,p} (-1)^{p-k}\binom{p}{k} W(p)t^k = \sum_p (t-1)^p W(p).\end{aligned}$$

Examples.

(1) The Euler Function

Given a positive integer N, what is the number $\phi(N)$ of integers k such that the greatest common divisor of k and N, $(k,N) = 1$ where $0 < k \leq N$?

Suppose that the distinct prime factors of N are $p_1,\ldots,p_n$. Then

$(k,N) = 1$ implies that k is not divisible by p_i, $i = 1,\ldots,n$, i.e., we ask for the number of integers not having the property of divisibility by p_i for every p_i. Let

α_1 be the condition k is divisible by p_1

α_2 " " " k " " " p_2

...

$\alpha_1\alpha_2$ " " " k " " " p_1p_2

etc. Then

$$\begin{aligned}\phi(N) &= N - \left(\frac{N}{p_1} + \frac{N}{p_2} + \cdots\right) + \left(\frac{N}{p_1p_2} + \frac{N}{p_1p_3} + \cdots\right) - \cdots \\ &= N\left(1 - \frac{1}{p_1}\right)\left(1 - \frac{1}{p_2}\right) \cdots \left(1 - \frac{1}{p_n}\right)\end{aligned}$$

the Euler function.

(2) Rencontres, Derangement or Montmort Problem

What is the number $R(n)$ of permutations P which change every element: $P(i) \neq i$ for $i = 1,\ldots,n$ (e.g., the card problem or "no one gets his own hat"...). Let α_i be the condition $P(i) = i$; then $\alpha_{i_1}\alpha_{i_2} \cdots \alpha_{i_s}$ is the condition $P(i_1) = i_1$, $P(i_2) = i_2,\ldots,P(i_s) = i_s$, clearly satisfied by $(n-s)!$ permutations, i.e., those which alter the remaining $n - s$. There are $\binom{n}{s}$ ways of choosing s conditions, and here the weight of each permutation is 1. Therefore,

$$\begin{aligned}E(0) = R(n) &= n! - \binom{n}{1}(n-1)! + \binom{n}{2}(n-2)! - \cdots + (-1)^n\binom{n}{n}(n-n)! \\ &= n!\left[1 - \frac{1}{1!} + \frac{1}{2!} - \frac{1}{3!} + \cdots + (-1)^n \frac{1}{n!}\right].\end{aligned}$$

$R(n)$ is the closest integer to $n!e^{-1}$ since the remainder in the series is less than $\frac{1}{n+1}$. To get $E(k)$ for $k > 0$, we use the generating function relation

$$\sum_k E(k)t^k = \sum_p (t-1)^p W(p) = \sum_{p=0}^{n} (t-1)^p \frac{n!}{p!} = n! \sum_{p=0}^{n} \frac{(t-1)^p}{p!} .$$

Now for large n this approaches $n!e^{t-1}$ and we conclude that $E(k)$ = coef of t^k in $n!e^{t-1} = \frac{n!}{k!} e^{-1}$.

(3) <u>The Menage Problem</u>

Count the number of permutations $U(n)$ of $1,\ldots,n$ for which neither $P(i) = i$ nor $P(i) = i + 1$ (modulo n).

(The number of ways of seating n married couples at a round table, with men and women alternating, such that no husband sits next to his wife is the product of $2(n!)$ for placing the women and $U(n)$ for avoiding couples.) There are now $2n$ conditions to consider.

Let α_{2i} be the condition $P(i) = i$

α_{2i+1} be the condition $P(i) = i + 1 \pmod n$.

The number of permutations satisfying k conditions is again $(n-k)!$. But the k conditions must be consistent (if this requirement is dropped there are $\binom{2n}{k}$ ways of choosing the conditions). The conditions are inconsistent if α_{2i+1}: $P(i) = i + 1$ and α_{2i}: $P(i) = i$ or α_{2i-1}: $P(i-1) = i$ and α_{2i}: $P(i) = i$ (i.e., a man's place is either unoccupied or occupied twice. Therefore, for consistency no two adjacent conditions can be allowed.

Let $v_k(N)$ be the number of ways of choosing k elements (conditions) on a circle of N elements such that no two are adjacent. Then

$$U(n) = \sum_k (-1)^k (n-k)! v_k(2n).$$

Let us compute $v_k(N)$. From the diagram

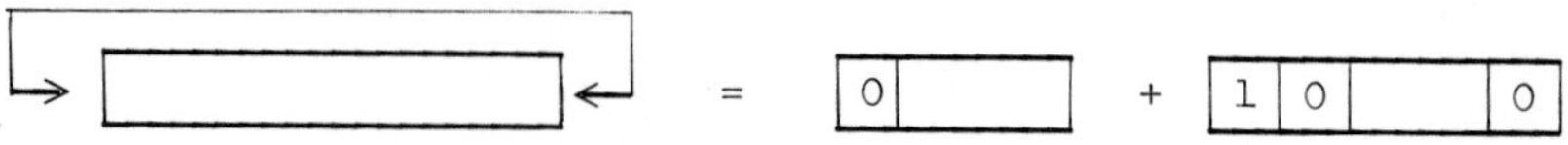

we have

$$v_k(N) = u_k(N-1) + u_{k-1}(N-3)$$

where $u_k(N)$ is the number of ways of choosing k non-adjacent elements on a line of N elements. But again diagrammatically

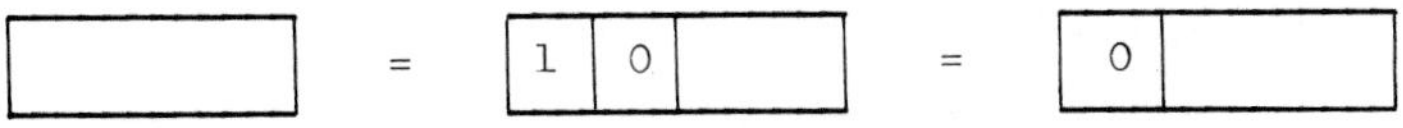

$$u_k(N) = u_{k-1}(N-2) + u_k(N-1)$$

As boundary conditions we clearly have

$$u_0(N) = 1, \qquad u_1(N) = N.$$

Let $u_k(N) = u_{k,N-k}$. Then for $N - k = p$ the above becomes

$$u_{k,p} = u_{k-1,p-1} + u_{k,p-1}, \quad u_{0,p} = 1, \quad u_{1,p} = p + 1.$$

This is the recursion relation for binomial coefficients $u_{k,p} = \binom{p+a}{k+b}$ so that using the boundary conditions we have

$$u_{k,p} = \binom{p+1}{k} \qquad \text{or} \qquad u_k(N) = \binom{N-k+1}{k}.$$

Hence

$$v_k(N) = u_k(N-1) + u_{k-1}(N-3) = \frac{N}{N-k}\binom{N-k}{k}$$

and

$$U(n) = \sum_{k=0}^{\infty} (-1)^k (n-k)! \frac{2n}{2n-k} \binom{2n-k}{k}.$$

2§. Permutations with Restricted Position. The Master Theorem

We recall that $\operatorname{Per}\begin{pmatrix} a & a & \cdots \\ b & b & \cdots \\ c & c & \cdots \\ \cdots \end{pmatrix}$ enumerates the ordered subsets of $S = \{a,b,c,\ldots\}$, where $\operatorname{Per}(a_{ij}) = \sum_P a_{P(1),1} a_{P(2),2} \cdots a_{P(n),n}$, and because of linearity, in row and column vectors, the permanent may be expanded by row, column or diagonal as with determinants. But no other tricks are allowed. Hence recursion relations are easy to get for structured matrices, but little else. To count permutations in which each element i is prohibited from moving to certain locations $\alpha_i, \beta_i, \gamma_i \cdots$ is now direct: number of permutations with prohibited positions $= \operatorname{Per}(a_{ij})$, where $a_{ij} = \begin{cases} 1 & \text{if } i \to j \text{ is allowed,} \\ 0 & \text{otherwise.} \end{cases}$ Thus, for rencontres,

$$R(n) = \operatorname{Per}\begin{pmatrix} 0 & 1 & \cdots & 1 \\ 1 & 0 & 1 & 1 \\ 1 & 1 & 0 & 1 \\ \cdots & & \ddots & \\ 1 & & 1 & 0 \end{pmatrix} = \operatorname{Per}(11'-I)$$

where $1' = (1,\ldots,1)$, $1 = \begin{pmatrix} 1 \\ \vdots \\ 1 \end{pmatrix}$ and I is the unit matrix. Similarly for menage

$$U(n) = \operatorname{Per}\begin{pmatrix} 0 & 0 & 1 & 1 & \cdots & 1 \\ 1 & 0 & 0 & 1 & & 1 \\ & 1 & & \ddots & & \vdots \\ \vdots & & & 0 & 0 & 1 \\ 1 & 1 & & 1 & 0 & 0 \\ 0 & 1 & \cdots & 1 & 1 & 0 \end{pmatrix}$$

$$= \operatorname{Per}(11'-I-P)$$

where

$$\text{Per} = \begin{pmatrix} 0 & 1 & \cdots & 0 \\ \vdots & & \ddots & \vdots \\ 0 & 0 & \ddots & 1 \\ 1 & 0 & \cdots & 0 \end{pmatrix}.$$

Let us look at a direct permanent expansion of $R(n)$. Expanding by the bottom row we get

$$R(n) = (n-1)\ S(n-1)$$

where

$$S(n) = \text{Per} \begin{pmatrix} 1 & 1 & \ddots & & & 1 \\ 0 & 1 & \ddots & \ddots & & \vdots \\ 1 & 0 & \ddots & \ddots & \ddots & \vdots \\ \vdots & & \ddots & \ddots & 1 & 1 \\ 1 & & 1 & \ddots & 0 & 1 \end{pmatrix}.$$

On the other hand expanding by the first row

$$S(n) = R(n-1) + (n-1)S(n-1).$$

Hence we have the recursion relation

$$R(n+1) = n\ R(n) + n\ R(n-1)$$

which is solvable in many ways.

<u>Exercises.</u>

1) From the generating function for the binomial coefficients $\binom{n}{r}$, show

that

$$\sum_{r=0}^{n} \binom{n}{r}^2 = \binom{2n}{n}.$$

2) The probability that a random walk starting at the origin of a one-dimensional space returns to the origin for the yth time at the $2n^{th}$ step is given by

$$P_y(2n) = \frac{y}{2n-y} \binom{2n-y}{n} 2^{y-2n}.$$

By summing over y, find the probability that the walk returns to the origin at the $2n^{th}$ step.

3) A set consists of n distinct elements. A sample of size m is chosen. Show that the probability that every element is selected at least once in d samplings (elements returned after each sample) is

$$P = 1 - nb_1^d + \binom{n}{2} b_2^d + \cdots + (-1)^n b_n^d$$

where

$$b_s \equiv \binom{n-s}{m} / \binom{n}{m}.$$

4) Solve the menage problems by direct permanent expansion.

An alternative representation of the permanent which leads to a rather general computational scheme is the multivariable generating function

$$\text{Per } (a_{ij}) = \text{coef of } x_1 \ldots x_n \text{ in } \left(\sum_{j=1}^{n} a_{1j} x_j \right) \left(\sum_{j=1}^{n} a_{2j} x_j \right) \cdots \left(\sum_{j=1}^{n} a_{nj} x_j \right)$$

so that for example:

the number of permutations with prohibited position = $P(a_{ij})$

$$= \text{coef of } x_1 \ldots x_n \text{ in } \prod_{i=1}^{n} \sum_{j=1}^{n} a_{ij}x_j$$

where

$$a_{ij} = \begin{cases} 1 & \text{if } i \to j \text{ is allowed} \\ 0 & \text{otherwise.} \end{cases}$$

As a painless generalization we may define:

$$\operatorname{Per}^{(n_1..n_s)}(a_{ij}) \equiv \text{coef of } x_1^{n_1} \ldots x_s^{n_s} \text{ in } \left(\sum_{j=1}^{s} a_{1j}x_j\right)^{n_1} \cdots \left(\sum_{j=1}^{s} a_{sj}x_j\right)^{n_s}$$

with the basic relation: the ordered permutations of $n_1 a_1$'s, $n_2 a_2$'s,... are given by:

$$\operatorname{Per}^{(n_1..n_s)} \begin{pmatrix} a_1 & a_1 & \cdots & a_1 \\ a_2 & a_2 & & \vdots \\ \vdots & & & \\ a_s & a_s & \cdots & a_s \end{pmatrix} = \text{coef of } x_1^{n_1} \ldots x_s^{n_s} \text{ in } \left(\sum_{j=1}^{s} a_j x_j\right)^{\sum_{i=1}^{s} n_i} .$$

Now for a systematic computation, we may recast the problem in a determinatal form, obtaining the so-called Master Theorem of MacMahon. To do this we have

$$\sum_{n_1..n_s} z_1^{n_1} \ldots z_s^{n_s} \text{ coef of } x_1^{n_1} \ldots x_s^{n_s} \text{ in } f_1(\underset{\sim}{x})^{n_1} \ldots f_s(\underset{\sim}{x})^{n_s}$$

$$= \sum_{n_1 \ldots n_s} \frac{z_1^{n_1} \ldots z_s^{n_s}}{n_1! \ldots n_s!} \left(\frac{\partial}{\partial x_1}\right)^{n_1} \cdots \left(\frac{\partial}{\partial x_s}\right)^{n_s} f_1(\underset{\sim}{x})^{n_1} \cdots f_s(\underset{\sim}{x})^{n_s}\Bigg|_{\underset{\sim}{x}=\underset{\sim}{0}} .$$

But for any h satisfying suitable boundary conditions

$$\int\cdots\int\cdots\int h(\underset{\sim}{y}) \sum_{n_1..n_s} \frac{z_1^{n_1}\cdots z_s^{n_s}}{n_1!\ldots n_s!} \left(\frac{\partial}{\partial y_1}\right)^{n_1} \cdots \left(\frac{\partial}{\partial y_s}\right)^{n_s} f_1(\underset{\sim}{y})^{n_1} .. f_s(\underset{\sim}{y})^{n_s} (dy)^s$$

$$= \int\cdots\int\cdots\int \sum_{n_1..n_s} \frac{f_1(\underset{\sim}{y})^{n_1} .. f_s(\underset{\sim}{y})^{n_s} z_1^{n_1} .. z_s^{n_s}}{n_1! \ldots n_s!} \left(\frac{-\partial}{\partial y_1}\right)^{n_1} \cdots \left(\frac{-\partial}{\partial y_s}\right)^{n_s} h(\underset{\sim}{y})(dy)^s$$

$$= \int..\int..\int h(y_1 - z_1 f_1(\underset{\sim}{y}),\ldots,y_s - z_s f_s(\underset{\sim}{y}))(dy)^s$$

$$= \int..\int..\int h(\underset{\sim}{x}) J_z(\underset{\sim}{y}/\underset{\sim}{x})(dx)^s, \text{ where } x_i = y_i - z_i f_i(\underset{\sim}{y}).$$

and $J_z(\underset{\sim}{y}/\underset{\sim}{x})$ is the Jacobian of the transformation.

Now equating coefficients of $h(\underset{\sim}{x})$ in first and last lines (i.e., taking the variational derivative with respect to $h(\underset{\sim}{x})$) we get

$$\sum_{n_1..n_s} \frac{z_1^{n_1} \cdots z_s^{n_s}}{n_1! \cdots n_s!} \left(\frac{\partial}{\partial x_1}\right)^{n_1} \cdots \left(\frac{\partial}{\partial x_s}\right)^{n_s} f_1(\underset{\sim}{x})^{n_1} \cdots f_s(\underset{\sim}{x})^{n_s} = J_z(\underset{\sim}{y}/\underset{\sim}{x}).$$

Suppose $f_i(0,\ldots,0) = 0$ for every i; then $\underset{\sim}{y} = \underset{\sim}{0}$ implies $\underset{\sim}{x} = \underset{\sim}{0}$. Furthermore,

$\dfrac{\partial x_i}{\partial y_j} = \delta_{ij} - z_i \dfrac{\partial f_i(\underset{\sim}{y})}{\partial y_j}$, so that

$$\sum_{n_1\ldots n_s} \frac{z_1^{n_1} \cdots z_s^{n_s}}{n_1! \cdots n_s!} \left(\frac{\partial}{\partial x_1}\right)^{n_1} \cdots \left(\frac{\partial}{\partial x_s}\right)^{n_s} f_1(\underset{\sim}{x})^{n_1} \cdots f_s(\underset{\sim}{x})^{n_s}\Bigg|_{\underset{\sim}{x}=0}$$

$$= 1/\mathrm{Det}\left(\delta_{ij} - z_i \frac{\partial f_i(\underset{\sim}{y})}{\partial y_j}\right)\Bigg|_{\underset{\sim}{y}=0} .$$

Finally choosing $f_i(\underset{\sim}{x}) = \sum_j a_{ij}x_j$ we conclude that

$$\text{Per}^{(n_1..n_s)}(a_{ij}) = \text{coef of } x_1^{n_1} \ldots x_s^{n_s} \text{ in } \left(\sum_j a_{ij}x_j\right)^{n_1} \cdots \left(\sum_j a_{ij}x_j\right)^{n_s}$$

$$= \text{coef of } x_1^{n_1} \ldots x_s^{n_s} \text{ in } \frac{1}{\text{Det}\,(\delta_{ij} - x_i a_{ij})}$$

or

$$\text{Per}^{(n_1..n_s)}(a_{ij}) = \text{coef of } x_1^{n_1} \ldots x_s^{n_s} \text{ in } \frac{1}{\text{Det}\,(I-XA)}$$

where X is the diagonal matrix with elements $x_{ij} = x_i\delta_{ij}$. This is MacMahon's Master Theorem.

[Note, in precisely the same way, one can prove the many dimensional extensions of the Lagrange implicit function theorem: if

$$y_i = x_i + f_i(\underset{\sim}{y})$$

then

$$G(\underset{\sim}{y}) = \overline{G}(\underset{\sim}{x}) = \sum \frac{1}{n_1! \, .. \, n_s!} \left(\frac{\partial}{\partial x_1}\right)^{n_1} \cdots \left(\frac{\partial}{\partial x_s}\right)^{n_s}$$

$$f_1(\underset{\sim}{x})^{n_1} \ldots f_s(\underset{\sim}{x})^{n_s} \, G(\underset{\sim}{x}) \, \text{Det}\left(\delta_{ij} - \frac{\partial f_i(\underset{\sim}{x})}{\partial x_j}\right),$$

where $G(\underset{\sim}{y})$ is an arbitrary function of $\underset{\sim}{y}$ and $G(\underset{\sim}{x})$ is $\overline{G}(\underset{\sim}{y}(\underset{\sim}{x}))$ expressed in terms of $\underset{\sim}{x}$.]

The major virtue of the Master Theorem is that it will enable conversion of the permanent to a solvable partition problem in many cases.

Examples.

(1) How many permutations $A_s(\xi)$ of $a_1^\xi \ldots a_s^\xi$ are there such that a_i goes over to either a_i or a_{i+1} (mod s)? If $\xi = 1$ the answer is 2 since a_i either stays in its position or moves and pushes the others. The general answer is, by the Master Theorem,

$$A_s(\xi) = \mathrm{Per}^{(\xi,..\xi)}\begin{pmatrix} 1 & 1 & & & \\ & 1 & 1 & & \bigcirc \\ & & 1 & 1 & \\ & \bigcirc & & \ddots & \ddots & 1 \\ 1 & & & & 1 \end{pmatrix}$$

$$= \text{coef of } x_1^\xi .. x_s^\xi \text{ in } \begin{vmatrix} 1-x_1 & -x_1 & & \bigcirc \\ & 1-x_2 & \ddots & -x_{s-1} \\ -x_s & \bigcirc & \ddots & 1-x_s \end{vmatrix}^{-1} ;$$

now expanding by the first column we find

$$A_s(\xi) = \text{coef of } x_1^\xi \ldots x_s^\xi \text{ in } [(1-x_1) \ldots (1-x_s) - x_1x_2 \ldots x_s]^{-1}$$

$$= \text{coef of } x_1^\xi \ldots x_s^\xi \text{ in } [(1-x_1) .. (1-x_s)]^{-1} \left(1 - \frac{x_1 \ldots x_s}{(1-x_1) \ldots (1-x_s)}\right)^{-1}$$

$$= \text{coef of } x_1^\xi \ldots x_s^\xi \text{ in } \sum_{p=0}^{\infty} (x_1..x_s)^p[(1-x_1)(1-x_2) .. (1-x_s)]^{-(p+1)}$$

$$= \text{coef of } x_1^\xi \ldots x_s^\xi \text{ in } \sum_{p=0}^{\infty} \prod_{i=1}^{s} x_i^p(1-x_i)^{-(p+1)} ;$$

but the coef of x^ξ in $x^p(1-x)^{-(p+1)}$ equals $\binom{\xi}{p}$. Hence

$$A_s(\xi) = \sum_{p=0}^{\xi} \binom{\xi}{p}^s .$$

(2) Rencontre Problem

From the Master Theorem we can rephrase this as

$$R(n) = \text{coef of } x_1 \ldots x_n \text{ in } \operatorname{Det}(I-XA)^{-1}$$

where

$$A = \begin{pmatrix} 0 & 1 & \cdots & 1 \\ 1 & 0 & \ddots & \vdots \\ \vdots & & 0 & 1 \\ 1 & \cdots & 1 & 0 \end{pmatrix} \quad \text{and} \quad X = \begin{pmatrix} x_1 & & \bigcirc \\ & \ddots & \\ \bigcirc & & x_n \end{pmatrix}$$

Now

$$\operatorname{Det}(I-XA) = \operatorname{Det}(X^{-1}-A) \operatorname{Det} X = \prod_{i=1}^{n} x_i \operatorname{Det}(X^{-1}-A).$$

To compute $\operatorname{Det}(X^{-1}-A)$ we find the product of the eigenvalues of $(X^{-1}-A)$.

If $(X^{-1}-A)u = \lambda u$ then taking the ith component we have

$$\frac{u_i}{x_i} + u_i - \sum_{j=1}^{n} u_j = \lambda u_i$$

or solving for u_i we get

$$u_i = \frac{\sum_{j=1}^{n} u_j}{1 - \lambda + \frac{1}{x_i}} .$$

By summing over i and cancelling $\sum_{j=1}^{n} u_j$ we find the characteristic equation

$$1 = \sum_{i=1}^{n} \frac{1}{1 - \lambda + \frac{1}{x_i}} .$$

Multiplying by $\prod_{i=1}^{n} (1 - \lambda + \frac{1}{x_i})$ we get:

$$\prod_{i=1}^{n} (1 + \frac{1}{x_i} - \lambda) - \sum_{j=1}^{n} \prod_{i \neq j} (1 + \frac{1}{x_i} - \lambda) = 0.$$

Retaining only largest and smallest powers of λ this becomes:

$$(-\lambda)^n + \cdots + \prod_{i=1}^{n} (1 + \frac{1}{x_i}) - \sum_{j=1}^{n} \prod_{i \neq j} (1 + \frac{1}{x_i}) = 0.$$

Hence the product of the eigenvalues is the constant term which can be rewritten as

$$\prod_{i=1}^{n} (1 + \frac{1}{x_i}) \left(1 - \sum_{j=1}^{n} \frac{1}{1 + \frac{1}{x_j}} \right) = \text{Det } (X^{-1} - A).$$

Therefore,

$$R(n) = \text{coef of } x_1 \, .. \, x_n \text{ in } \prod_{i=1}^{n} (x_i + 1)^{-1} \left(1 - \sum_{j=1}^{n} \frac{x_j}{1 + x_j} \right)^{-1} .$$

Since we can neglect all powers of x_i above 1 then we may replace $(x_i + 1)^{-1}$ by e^{-x_i} and $\frac{x_j}{1 + x_j}$ by x_j. We conclude that

$$R(n) = \text{coef of } x_1 \ldots x_n \text{ in } e^{-\sum_{j=1}^{n} x_j} \left(1 - \sum_{j=1}^{n} x_j \right)^{-1} .$$

(i) Direct evaluation: expand each factor in power series

$$R(n) = \text{coef of } x_1 \ldots x_n \text{ in } \sum_{s=0}^{\infty} \sum_{t=0}^{\infty} \frac{(-1)^s}{s!} \left(\sum_{j=1}^{n} x_j \right)^{s+t}$$

$$= \sum_{s+t=n} \frac{(-1)^s}{s!} (s+t)! = n! \sum_{s=0}^{n} \frac{(-1)^s}{s!}$$

(ii) rationalization: since $\frac{1}{x} = \int_0^{\infty} e^{-\gamma x} d\gamma$. Then

$$R(n) = \text{coef of } x_1 \ldots x_n \text{ in } e^{-\sum_{j=1}^{n} x_j} \int_0^{\infty} e^{-\gamma} e^{\gamma \sum_{j=1}^{n} x_j} d\gamma$$

$$= \int_0^{\infty} e^{-\gamma} (\gamma - 1)^n d\gamma = e^{-1} \Gamma(n+1, -1)$$

The incomplete Γ function which is known to have the value $\sum_{s=0}^{n} (-1)^s \frac{n!}{s!}$.

(iii) generating function after rationalization:

$$\sum_{n=0}^{\infty} R(n) \frac{t^n}{n!} = \int_0^{\infty} e^{-\gamma} \sum_{n=0}^{\infty} (\gamma - 1)^n \frac{t^n}{n!} d\gamma$$

$$= \int_0^{\infty} e^{-\gamma} e^{t(\gamma-1)} d\gamma = \frac{e^{-t}}{1-t} = \sum_{a,b=0}^{\infty} \frac{(-1)^a t^a}{a!} t^b .$$

Hence equating the coefficients of t^n we get

$$\frac{R(n)}{n!} = \sum_{a=0}^{n} \frac{(-1)^a}{a!} .$$

(3) Menage Problem

Here $A = (a_{ij})$ where $a_{ij} = \begin{cases} 0 & \text{if } i = j \text{ or } i = j - 1 \pmod{n} \\ 1 & \text{otherwise} \end{cases}$

$$X = \begin{pmatrix} x_1 & & \\ & \ddots & \\ & & x_n \end{pmatrix} .$$

To proceed systematically, we decompose the matrix $A = 11' - I - P$ into an easily evaluated and inverted part plus a "separable" low rank part. Thus

$$A = 11' - \begin{pmatrix} 0 \\ 0 \\ \cdots \\ 0 \\ 1 \end{pmatrix} (1\ 0..0) - (I+E) \quad \text{where}$$

$$E = \begin{pmatrix} 0 & 1 & & \bigcirc \\ & \ddots & \ddots & \\ & & \ddots & 1 \\ \bigcirc & & & 0 \end{pmatrix}$$

Here $(I+E)$ is in fact upper triangular.

Now in general, for nonsingular M,

$$\text{Det}\left(M + \sum_{i=1}^{\gamma} u_i v_i'\right) = \text{Det } M \left(I + \sum_{i=1}^{\gamma} M^{-1} u_i v_i'\right) = \text{Det } M \text{ Det}\left(I + \sum_{i=1}^{\gamma} M^{-1} u_i v_i'\right) .$$

Let e_i be the ith unit vector, then

$$v_i = Ve_i \qquad \text{where} \qquad V = (v_1, v_2, \ldots, v_\gamma, e_{\gamma+1}, \ldots, e_n).$$

Then

$$\begin{aligned} \text{Det}\left(M + \sum_{i=1}^{\gamma} u_i v_i'\right) &= \text{Det }(M) \text{ Det}\left(I + \sum_{i=1}^{\gamma} M^{-1} u_i e_i' V'\right) \\ &= \text{Det }(M) \text{ Det}\left(I + \sum_{i=1}^{\gamma} V' M^{-1} u_i e_i'\right) \end{aligned}$$

$$= \text{Det}(M)\,\text{Det}\,[I + (V'M^{-1}u_1,\ V'M^{-1}u_2,\ldots,V'M^{-1}u_\gamma,0,\ldots,0)]$$

$$= \text{Det}(M)\,\text{Det}\,(\delta_{ij}+v_i'M^{-1}u_j)_{\gamma\times\gamma}\,.$$

We conclude that

$$\text{Det}\left(I - X\left(B + \sum_{i=1}^{\gamma} u_i v_i'\right)\right) = \text{Det}\,(I-XB)\ \text{Det}\,(\delta_{ij} - v_i'(I-XB)^{-1}Xu_j)_{\gamma\times\gamma}\,.$$

In the present case setting $X = -X$ to reduce the number of minus signs we get:

$$U(n) = (-1)^n \text{ coef of } x_1 \ldots x_n \text{ in}$$

$$\left[\text{Det}\,(I - X(I+E))\ \text{Det}\,(\delta_{ij} + v_i'(I - X(I+E))^{-1}Xu_j)_{2\times 2}\right]^{-1}$$

where

$$u_1 = \begin{pmatrix} 1\\ 1\\ 1\\ 1 \end{pmatrix},\ u_2 = -\begin{pmatrix} 0\\ \vdots\\ 0\\ 1 \end{pmatrix},\quad v_1 = \begin{pmatrix} 1\\ 1\\ \vdots\\ 1 \end{pmatrix},\quad v_2 = \begin{pmatrix} 1\\ 0\\ \vdots\\ 0 \end{pmatrix}$$

But

$$\text{Det}\,(I - X(I+E)) = \begin{vmatrix} 1-x_1 & -x_1 & & 0 \\ & \ddots & \ddots & \\ & & \ddots & -x_{n-1} \\ 0 & & & 1-x_n \end{vmatrix} = \prod_{i=1}^{n}(1-x_i)$$

and neglecting powers of x_i greater than 1:

$$(I-X-XE)^{-1}X = (I - \frac{X}{I-X}E)^{-1} \longrightarrow (I-XE)^{-1}X = X + XEX + XEXEX + \cdots\,.$$

Hence

$$[(I-XE)^{-1}X]_{ij} = x_i x_{i+1} \cdots x_j, \qquad i \le j,$$

and

$$\mathrm{Det}\,(\delta_{ij}+v_i'(I-X(I+E))^{-1}Xu_j)_{2\times 2} = \begin{vmatrix} 1+\sum\limits_{i\le j} x_i \cdots x_j & \sum\limits_j x_1 \cdots x_j \\ -\sum\limits_i x_i \cdots x_n & 1 - x_1 \cdots x_n \end{vmatrix}$$

$$= 1 - x_1 \cdots x_n + \sum_{i\le j} x_i \cdots x_j + \sum_{j<i} x_1 \cdots x_j\, x_i \cdots x_n .$$

Therefore,

$$U(n) = (-1)^n \Bigg[1 + \text{coef of } x_1 \cdots x_n$$

$$\text{in } \prod_{i=1}^{n} (1-x_i)^{-1} \Big(1 + \sum_{i\le j} x_i \cdots x_j + \sum_{j<i} x_1 \cdots x_j \; x_i \cdots x_n \Big)^{-1} \Bigg].$$

Now

$$(-1)^n \prod_{i=1}^{n} (1-x_i)^{-1} \Big(1 + \sum_{i\le j} x_i \cdots x_j + \sum_{j<i} x_1 \cdots x_j \; x_i \cdots x_n \Big)^{-1}$$

$$\to (-1)^n e^{\Sigma x_i} \int_0^\infty e^{-\gamma} e^{-\gamma \sum\limits_{i\le j} x_i \cdots x_j - \gamma \sum\limits_{j<i} x_1 \cdots x_j x_i \cdots x_n} d\gamma$$

$$= (-1)^n \int_0^\infty e^{\sum\limits_{i=1}^{n} (1-\gamma) x_i} e^{-\gamma \sum\limits_{i<j} x_i \cdots x_j} e^{-\gamma \sum\limits_{j<i} x_1 \cdots x_j x_i \cdots x_n} e^{-\gamma} d\gamma$$

$$\to (-1)^n \int_0^\infty \prod_{i=0}^{n} (1 + (1-\gamma)x_i) \prod_{i<j} (1-\gamma x_i \cdots x_j) \Big(1 - \gamma \sum_{j<i} x_1 \cdots x_j x_i \cdots x_n \Big) e^{-\gamma} d\gamma .$$

We wish to find the coefficient of $x_1 \ldots x_n$ in the above expression. This is equivalent to a weighted partition problem, that of decomposing the set $(x_1,\ldots,x_n)$ into the subsets which are present in the integrand and weighting each by its proper function of γ.

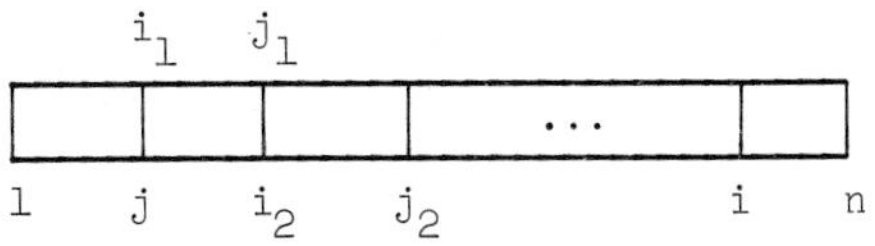

The subsets present in the integral are of the form x_i, $x_i \ldots x_j$ for $i < j$, and $x_1 \ldots x_j \; x_i \ldots x_n$ for $j < i$. The counting problem can be converted immediately to a one variable generating function: we represent the above subsets by x, x^{j+1-i}, and $x^{n+1-i+j}$ respectively and restrict their combinations to the set $x_1 \ldots x_n$ by choosing only the coefficient of x^n. In the diagram the left-right weight is clearly

$$1 - \gamma(x+x^2+\ldots)(x+x^2+x^3+\ldots) = 1 - \frac{\gamma x^2}{(1-x)^2}$$

whereas each intermediate part has a weight of

$$(1-\gamma)x - \gamma(x^2+x^3+\cdots) = (1-\gamma)x - \frac{\gamma x^2}{1-x} = x - \frac{\gamma x}{1-x} .$$

Thus a weight of an arbitrary number of intermediate parts is $1 + (x - \frac{\gamma x}{1-x}) + (x - \frac{\gamma x}{1-x})^2 + \cdots = [1 - x + \frac{\gamma x}{1-x}]^{-1}$. Hence the total weight is

$$\left[1 - \frac{\gamma x^2}{(1-x)^2}\right] [1 - x + \frac{\gamma x}{1-x}]^{-1} = \frac{-x}{1-x} + \frac{1+x}{1-x} \frac{1}{1 + \gamma \frac{x}{(1-x)^2}}$$

and, therefore,

$$U(n) = (-1)^n \left(1 + \text{coef of } x^n \text{ in } \frac{-x}{1-x} \int_0^\infty e^{-\gamma} d\gamma + \frac{1+x}{1-x} \int_0^\infty e^{-\gamma} \frac{d\gamma}{1 + \frac{\gamma x}{(1-x)^2}}\right)$$

$$= (-1)^n \left(\text{coef of } x^n \text{ in } \frac{1+x}{1-x} \int_0^\infty e^{-\gamma} \frac{d\gamma}{1 + \frac{\gamma x}{(1-x)^2}} \right) .$$

We thus have the generating function (letting $x \to -x$) $\sum_{n=0}^{\infty} U(n)x^n =$

$\frac{1-x}{1+x} \int_0^\infty e^{-\gamma} \frac{d\gamma}{1 - \frac{\gamma x}{(1+x)^2}}$ which is an exponential integral.

To evaluate $U(n)$ explicitly, we expand in γ:

$$\left[1 - \frac{\gamma x}{(1+x)^2} \right]^{-1} = \sum_{p=0}^{\infty} \gamma^p x^p (1+x)^{-2p}$$

so that

$$\frac{1-x}{1+x} \int_0^\infty e^{-\gamma} \frac{d\gamma}{1 - \frac{\gamma x}{(1+x)^2}} = \sum_{p=0}^{\infty} p!(1-x)x^p(1+x)^{-(2p+1)}$$

$$= \sum_{k=0}^{\infty} \sum_{p=0}^{\infty} p! \ (-1)^k [x^{p+k} \binom{2p+k}{k} - x^{p+k+1} \binom{2p+k}{k}]$$

$$= \sum_{n=0}^{\infty} x^n \left\{ \sum_{k=0}^{n} (-1)^k \binom{2n-k}{k} (n-k)! - \sum_{k=0}^{n-1} (-1)^k \binom{2n-k-2}{k} (n-k-1)! \right\}$$

Hence

$$U(n) = \sum_{k=0}^{n} (-1)^k [\binom{2n-k}{k} + \binom{2n-k-1}{k-1}](n-k)!$$

$$= \sum_{k=0}^{n} (-1)^k (n-k)! \binom{2n-k}{k} \frac{2n}{2n-k} .$$

3§. Extension of the Master Theorem

There is an important alternate representation of the determinant-permanent dualism which bypasses the Lagrange implicit function derivation and often permits determinantal representation of other counting processes as well. We observe that

$$\begin{aligned}
\operatorname{Per}(a_{ij}) &= \text{coef of } x_1 \dots x_n \text{ in } [\operatorname{Det}(I-XA)]^{-1} \\
&= \text{''} \quad [\operatorname{Det}(I-A'X)(I-XA)]^{-1/2} \\
&= \text{''} \quad [\operatorname{Det}(I-A'X-XA+A'X^2A)]^{-1/2} \\
&= \text{''} \quad [\operatorname{Det}(I-A'X-XA)]^{-1/2}.
\end{aligned}$$

But

$$(\operatorname{Det} B)^{-1/2} = \int\cdots\int e^{-\frac{1}{2}\Sigma y_i b_{ij} y_j} \left(\frac{dy}{\sqrt{2\pi}}\right)^n .$$

Hence

$$\begin{aligned}
\operatorname{Per}(a_{ij}) &= \text{coef of } x_1 \dots x_n \text{ in } \int\cdots\int e^{-\frac{1}{2}\Sigma y_i^2} e^{\Sigma y_i (XA)_{ij} y_j} \left(\frac{dy}{\sqrt{2\pi}}\right)^n \\
&= \text{''} \quad \int\cdots\int e^{-\frac{1}{2}\Sigma y_i^2} e^{\Sigma x_i y_i a_{ij} y_j} \left(\frac{dy}{\sqrt{2\pi}}\right)^n .
\end{aligned}$$

Observing that coef $x_1 \dots x_n$ in $e^{\Sigma x_i c_i} = \prod c_i$ this becomes

$$\operatorname{Per}(a_{ij}) = \int\cdots\int e^{-\frac{1}{2}\Sigma y_i^2} \prod_i y_i \prod_i \left(\sum_j a_{ij} y_j\right) \left(\frac{dy}{\sqrt{2\pi}}\right)^n .$$

Now $y_i e^{-\frac{1}{2} y_i^2}$ integrates to 0 if a given term in the product $\prod_i \left(\sum_j a_{ij} y_j\right)$ does not contain y_i; it integrates to 1 if multiplied by a single y_i from the product. Hence the only terms in the product which contribute are those which contain at least one power of y_i for each i. But the product is only of degree n. Therefore, only terms of the form $y_1 \ldots y_n$ in the product give a nonvanishing result. We conclude that

$$\text{Per}\ (a_{ij}) = \text{coef}\ y_1 \ldots y_n \text{ in } \prod_i \left(\sum_j a_{ij} y_j\right).$$

If this proof is presented in reverse order one gets the Master Theorem for permanents.

The importance of this formulation lies in the fact that although it is generally easy to convert a counting process to coefficient determination of a many variable generating function, this generating function need not have a form as simple as $\prod_i \left(\sum_j a_{ij} y_j\right)$. Nevertheless, the above derivation may often be used to rewrite the generating function as a simple determinant. As an example we consider the problem of the zero point entropy of two dimensional ice.

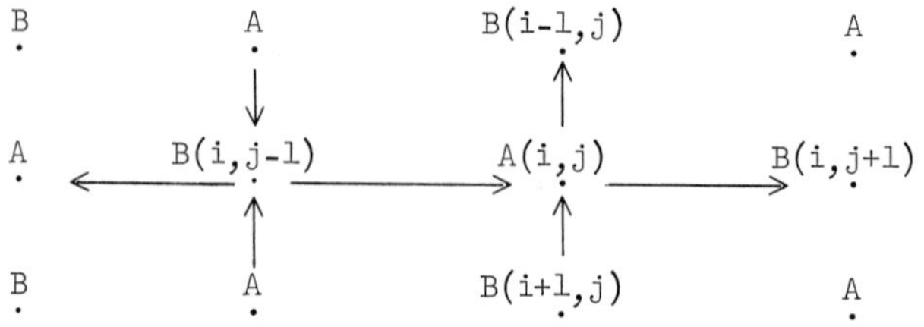

The precise meaning of this phrase will be spelled out in more detail later. At the moment we can just regard it as the problem of choosing a full set of directed bonds between vertices on a square lattice such that 2 bonds are directed into each vertex and two bonds out of each vertex. (This can be achieved only if periodic boundary conditions are assumed; otherwise the bonds on the boundary of the lattice must be specified.)

It is convenient to divide the vertices into A sites and B sites. An A site is one for which the sum of the row and column indices is even; a B site for which the sum is odd. Then our requirement is simply that each A-site sends out two arrows and each B-site receives 2 arrows. (Because a bond can only connect an A-site to a B-site). Let $I(M,N)$ be the number of ways arrows can be drawn on an $M \times N$ lattice such that each A-site has two arrows going out and each B-site two arrows coming in. An A-site has $\binom{4}{2}$ possible configurations of arrows; correspondingly, we can associate with it the weight

$$A_{ij} = x_{i-1,j}x_{i,j+1} + x_{i-1,j}x_{i+1,j} + x_{i-1,j}x_{i,j-1} + x_{i,j+1}x_{i+1,j} + x_{i,j+1}x_{i,j-1} + x_{i+1,j}x_{i,j-1},$$

where $x_{k,\ell}$ indicates an arrow to the B-site (k,ℓ). Thus each $x_{k,\ell}$ must occur twice and we have

$$I(M,N) = \text{coef of } \prod_{i+j \text{ odd}} x_{ij}^2 \text{ in } \prod_{i+j \text{ even}} A_{ij}.$$

We are thus led to the general problem of finding

$$P(a_{ijk}) = \text{coef of } z_1^2 \cdots z_n^2 \text{ in } \prod_{i=1}^{n} \Big(\sum_{j,k=1}^{n} a_{ijk} z_j z_k \Big).$$

We note that

$$\int_{-\infty}^{\infty} z^i \, \frac{z^2-1}{2} \, e^{-\frac{1}{2}z^2} \frac{dz}{\sqrt{2\pi}} = \begin{cases} 1 & i = 2 \\ 0 & i = 0,1 \end{cases}.$$

Hence $\prod_{i=1}^{n} \left(\frac{z_i^2 - 1}{2} e^{-\frac{1}{2}z_i^2} \right)$ must be multiplied by at least the second power of each z_i in order that the resulting integral not vanish. In particular, if it is multiplied by a multinomial of degree $2n$ in the z_i's then only the term $z_1^2 z_2^2 \cdots z_n^2$ will contribute to the integral. Therefore, we have

$$P(a_{ijk}) = \left(\frac{1}{2}\right)^n \int_{-\infty}^{\infty}\cdots\int e^{-\frac{1}{2}\sum_{i=1}^{n} z_i^2} \prod_{i=1}^{n} (z_i^2-1) \prod_{i=1}^{n} \left(\sum_{j,k} a_{ijk} z_j z_k\right) \left(\frac{dz}{\sqrt{2\pi}}\right)^n$$

$$= \text{coef of } x_1 \cdots x_n \; y_1 \cdots y_n \text{ in}$$

$$\int\cdots\int e^{-\frac{1}{2}\Sigma z_i^2} e^{\frac{1}{2}\Sigma x_i(z_i^2-1)} e^{\Sigma y_i a_{ijk} z_j z_k} \left(\frac{dz}{\sqrt{2\pi}}\right)^n$$

$$= \text{" " " "} \quad e^{-\frac{1}{2}\Sigma x_i} \left[\operatorname{Det}\left((1-x_j)\delta_{jk} - 2\sum_i y_i a_{ijk}\right)\right]^{-1/2}$$

$$= \text{" " " "} \quad e^{-\frac{1}{2}\Sigma x_i} \prod_i (1-x_i)^{-1/2} \operatorname{Det}\left[\delta_{jk} - 2\sum_i \frac{y_i}{1-x_j} a_{ijk}\right]^{-1/2}$$

$$= \text{" " " "} \quad \operatorname{Det}\left[\delta_{jk} - 2\sum_i (y_i + y_i x_j) a_{ijk}\right]^{-1/2}$$

$$= \text{" " " "} \quad \operatorname{Det}\left[\delta_{jk} - 2\sum_i y_i x_j a_{ijk}\right]^{-1/2},$$

where we have used the fact that $e^{-\frac{1}{2}x}(1-x)^{-\frac{1}{2}} \equiv 1 \pmod{x^2}$. In the above case of the ice model $n = \frac{NM}{2}$ and

$$A_{ij} = \sum_{r,s,t,u} a_{ij,rs,tu} x_{rs} x_{tu} .$$

It is often possible to rephrase a counting problem so that the Master Theorem is directly applicable. The ice problem is of this nature. Consider a bond between an A-site and a B-site. Let us give this bond a weight x_A if the arrow goes to A and x_B if the arrow goes to B. Since each site must receive exactly two arrows, we have at once

$$I(M,N) = \text{coef of } \prod_A x_A^2 \prod_B x_B^2 \text{ in } \prod_{(A,B)} (x_A + x_B),$$

where (A,B) designate a pair joined by a bond. Thus in general we have to compute

$$P_2(a_{ij}) = \text{coef of } \prod_{i=1}^{n} x_i^2 \text{ in } \prod_{i=1}^{2n} \sum_{j=1}^{n} a_{ij}x_j.$$

We can do this by pairing the $2n$ rows of $a = (a_{ij})$:

$$b_{ij} \equiv a_{n+i,j} ,$$

and noting that ab = coef of y in $(a + \frac{1}{2} by)^2$. In other words, we have immediately

$$P_2(a_{ij}) = \text{coef of } \prod_{i=1}^{n} y_i \text{ in coef } \prod_{i=1}^{n} x_i^2 \text{ in } \prod_{i=1}^{n} \left[\sum_{j=1}^{n} (a_{ij} + \frac{1}{2} y_i a_{n+i,j}) x_j \right]^2 ,$$

and the Master Theorem is applicable again.

C. Partitions, Compositions and Decompositions

1§. Permutation Counting As a Partition Problem

a) Counting with allowed transitions.

Let us investigate the genesis of the partition enumeration involved in the permutation counting problems that we have considered. Let us analyze

$$\text{Per } (a_{ij}) = \sum_{p} a_{P(1),1} \cdots a_{P(n),n}$$

in terms of allowed transitions. Any permutation P can be decomposed into cycles, e.g., if $P[1,2,3,4,5,6] = [1,5,6,2,4,3]$, we denote this by $P = (1)(2,5,4)(3,6)$, meaning that $1 \to 1$, $2 \to 5 \to 4 \to 2$, $3 \to 6 \to 3$. More generally, we can write $P = c_1 c_2 \ldots c_s$, where $c = (j_1, \ldots, j_h)$ is of order h, i.e , $c^h = I$, and the elements of $c_1, \ldots, c_s$ form a partition of $1, \ldots, n$. Now if c contains (acts on) q, the product $a_{q,c(q)} a_{c(q),c^2(q)} \cdots a_{c^{h-1}(q),q}$ must appear in the expansion of the permanent. Suppose we have a $0 - 1$ matrix, i.e., 1 for an allowed transition, 0 otherwise; then the above product will not vanish if every transition in the cycle is allowed. We conclude that

$$\text{Per } (a_{ij}) = \text{No. of decompositions of } 1,\ldots,n \text{ into allowed closed paths or cycles.}$$

<u>Example</u>.

How many permutations of n ordered elements do not allow any element to shift more than one position to the right? (Boundary not periodic.)

The allowed closed paths are clearly built up of sequences of adjacent elements of any length. We wish to find the number of ways that the n elements can be decomposed into such sequences.

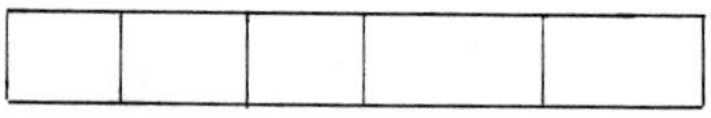

If each element has weight x then the aggregate weight of a single sequence is $x + x^2 + x^3 + \cdots$ according to the number of elements it can contain, or $\frac{x}{1-x}$. Similarly p sequences will have weight $(\frac{x}{1-x})^p$, and all possible sets of sequences weight $\frac{x}{1-x} + (\frac{x}{1-x})^2 + \cdots = \frac{x}{1-2x}$. We choose those sequences which combined have n elements by taking the coefficient of x^n; number $\#(n) =$ coef of x^n in $\frac{x}{1-2x} = 2^{n-1}$. [More directly, consider the leading element of the possible sequences, element 1 must lead and any subset of remaining $n - 1$ elements can lead; therefore $\#(n) = 2^{n-1}$.]

With periodic boundary the only alteration is that the first sequence (shaded) can start on the right. It is easily seen that the weight of the first sequence is therefore,

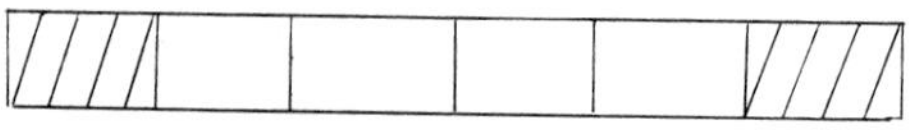

$x + 2x^2 + 3x^3 + \cdots = \frac{x}{(1-x)^2}$, except for an n-cycle, which is here counted n-times instead of once. Thus $\#(n) = -(n-1) +$ coef of x^n in $\frac{x}{(1-x)^2} \frac{1-x}{(1-2x)} = 2^n - n$.

b) <u>Counting with prohibited transitions</u>

If prohibited transitions are specified instead, the Inclusion-Exclusion theorem asks us to look at paths on subsets instead. We can see this quite directly from the permanent viewpoint. We have already noted that since the permanent is linear in rows and columns,

$$\text{Per }(A+B) = \sum_j \sum_{s_j, t_j} \text{Per }(A)_{[s_j, t_j]} \text{ Per }(B)_{[\bar{s}_j, \bar{t}_j]}$$

where $\dim s_j = \dim t_j = j$ and the row and column sets are in brackets. Thus

$$\begin{aligned}\text{Per }(11'-B) &= \sum_{j, s_j, t_j} \text{Per }(11')_{[\bar{s}_j, \bar{t}_j]} (-1)^j \text{ Per }(B)_{[s_j, t_j]} \\ &= \sum_{j, s_j, t_j} (-1)^j \ (n-j)! \text{ Per }(B)_{[s_j, t_j]} ,\end{aligned}$$

which is precisely the Inclusion-Exclusion theorem.

A different counting in terms of closed paths arises from the Master theorem. We recall from page 23 that

$$\text{Det }(I - X(11'-B)) = (\text{Det }(I+XB))(1 - \sum_{i,j} (I+XB)^{-1}_{ij} x_j),$$

so that

$$\text{Per }(11'-B) = \text{coef of } x_1 \cdots x_n \text{ in } [\text{Det }(I+XB)]^{-1}$$

$$\int_0^\infty e^{-\gamma} e^{\gamma \sum_{i,j} (I+XB)^{-1}_{ij} x_j} \, d\gamma,$$

or replacing X by $-X$ and expanding the inverse in the exponent,

$$\text{Per }(11'-B) = \text{coef of } x_1 \ldots x_n (-1)^n \text{ in}$$

$$\frac{1}{\text{Det }(I-XB)} \int_0^\infty e^{-\gamma} e^{-\gamma \sum_{i,j} (X+XBX+XBXBX+\cdots)_{ij}} \, d\gamma.$$

This is again an expression in terms of allowed paths, which now occur both in the determinant and in the exponent.

2§. Classification of Partitions

The general problem of distributing objects among classes is known as partitioning. Technically, we differentiate between the cases in which the objects are or are not distinguishable (labeled) and similarly for the classes. We also have to specify whether or not a class can be empty. The distribution of n distinguishable objects into M distinguishable classes is trivial -- there are M^n possibilities, allowing empty classes. The other designations that we shall use are given in the following table

	distinguishable objects	indistinguishable objects
distinguishable classes		composition
indistinguishable classes	decomposition	partition

(a) Distribution of unlabeled objects: compositions

The composition of n elements into m classes can also be interpreted as the subdivisions of n ordered objects into m connected groups. We have already considered this problem without restriction on the number m. If $c_{m,n}$ is the number of nonempty m-part compositions of n then attaching a weight t to each object, as before, the class weight is again $t + t^2 + \dots = \frac{t}{1-t}$, so that

$$\sum_{n=1}^{\infty} c_{m,n}\, t^n = \left(\frac{t}{1-t}\right)^m = \sum_{p=0}^{\infty} t^{m+p} \binom{m-1+p}{p}.$$

Hence

$$c_{m,n} = \binom{n-1}{n-m} = \binom{n-1}{m-1},$$

which is also the number of ways of selecting $m - 1$ class leaders out of the $n - 1$ elements remaining after the first class necessarily starting with element 1 has been chosen. If c_n is the total number of nonempty compositions without specifying m then

$$\sum_{n=1}^{\infty} c_n t^n = \sum_{m,n=1}^{\infty} c_{m,n} t^n = \sum_{m=1}^{\infty} \left(\frac{t}{1-t}\right)^m = \frac{t}{1-2t}$$

or

$$c_n = 2^{n-1}$$

as we saw previously.

If we restrict the classes to no more than s elements then clearly

$$\sum_{n=1}^{\infty} c_{m,n}^{(s)} t^n = (t+t^2+\cdots+t^2)^m = \left(\frac{t-t^{s+1}}{1-t}\right)^m$$

$$\sum_{n=1}^{\infty} c_n^{(s)} t^n = \sum_{m=1}^{\infty} \left(\frac{t-t^{s+1}}{1-t}\right)^m = \frac{t-t^{s+1}}{1-2t+t^{s+1}} .$$

Explicit evaluation of $c_n^{(s)}$ for $s > 2$ is complicated.

In statistical mechanics the problem of counting compositions with empty classes allowed occurs frequently. Suppose we wish to place n identical particles in m energy levels. Then

$$\sum_{n=1}^{\infty} \bar{c}_{m,n} t^n = (1+t+t^2+\cdots)^m = (1-t)^{-m}$$

hence

$$\bar{c}_{m,n} = \binom{m-1+n}{n}.$$

If the occupancy of each level is restricted to s particles, then

$$\sum_{n=1}^{\infty} \bar{c}_{m,n}^{(s)} t^n = (1+t+\cdots+t^s)^m = \left(\frac{1-t^{s+1}}{1-t}\right)^m .$$

In particular for Fermi-Dirac statistics, $s = 1$, yielding a generating function of $(1+t)^m$ and

$$\bar{c}_{m,n}^{(1)} = \binom{n}{m}.$$

(b) Distribution of unlabeled objects: partitions

Let Π_p^n be the number of partitions of n into p non-empty classes, that is, the number of ways n unlabeled (indistinguishable) object can be distributed among p indistinguishable non-empty boxes. We may represent a partition by its canonical form:

$$n = a_1 + a_2 + \cdots + a_p \qquad \text{where} \qquad a_1 \geq a_2 \geq \cdots \geq a_p > 0.$$

It is easy to derive a recursion relation for Π_p^n. We do this by constructing a partition in the following sequence. First place one object in each box, then distribute the remaining $n - p$ objects in those k boxes which originally contained more than one object. Clearly $1 \leq k \leq p$, and so we have

$$\Pi_p^n = \Pi_0^{n-p} + \Pi_1^{n-p} + \cdots + \Pi_p^{n-p} .$$

If the same relation is applied to Π_{p-1}^{n-1} then

$$\Pi_{p-1}^{n-1} = \Pi_0^{n-p} + \Pi_1^{n-p} + \cdots + \Pi_{p-1}^{n-p}$$

whence

$$\Pi_p^n = \Pi_{p-1}^{n-1} + \Pi_p^{n-p} \quad .$$

This relation can be used to generate the complete set of partition numbers since, on iteration, one either arrives at $\Pi_0^q = \begin{cases} 1, & q = 0 \\ 0, & q > 0 \end{cases}$, or $\Pi_p^q = 0$ for $q < p$.

A more convenient formal expression for the partition numbers arises via the generating function

$$f(x,y) = \sum_{n=0}^{\infty} \sum_{p=0}^{\infty} \Pi_p^{n+p} \; x^n y^p .$$

To find $f(x,y)$ we take the recursion relation $\Pi_p^{n+p} = \Pi_{p-1}^{n+p-1} + \Pi_p^n$ and so obtain

$$\sum_{n=0}^{\infty} \sum_{p=0}^{\infty} \Pi_p^{n+p} \; x^n y^p = \sum_{n=0}^{\infty} \sum_{p=0}^{\infty} \Pi_{p-1}^{n+p-1} \; x^n y^p + \sum_{n=0}^{\infty} \sum_{p=0}^{\infty} \Pi_p^n \; x^n y^p \; ,$$

which is the functional equation

$$f(x,y) = yf(x,y) + f(x,xy).$$

Hence

$$f(x,y) = \frac{1}{1-y} f(x,xy) = \frac{1}{1-y} \; \frac{f(x,x^2y)}{1 - xy} = \cdots$$

$$= \frac{1}{1-y} \; \frac{1}{1-xy} \; \frac{1}{1-x^2y} \; \cdots \; \frac{1}{1-x^s y} \; f(x,x^{s+1}y).$$

Now for $|x| < 1$ and $s \to \infty$ we get $f(x,x^{s+1}y) \to f(x,0) = \sum_{n=0}^{\infty} \Pi_0^n x^n = \Pi_0^0 x^0 = 1$

and we conclude that

$$f(x,y) = \prod_{s=0}^{\infty} \frac{1}{(1-x^s y)} .$$

As an immediate consequence we can find the generating function for the total partition number,

$$\Pi^n = \sum_{p=0}^{n} \Pi_p^n .$$

We need only observe that $f(x,x) = \sum_{n=0}^{\infty} \sum_{p=0}^{\infty} \Pi_p^{n+p} x^{n+p} = \sum_{m=0}^{\infty} \sum_{p=0}^{m} \Pi_p^m x^m = \sum_{m=0}^{\infty} \Pi^m x^m$.

In other words

$$\sum_{m=0}^{\infty} \Pi^m x^m = f(x,x) = \prod_{s=1}^{\infty} \frac{1}{(1-x^s)} .$$

It is illuminating to see why this must be true. We expand out each reciprocal:

$$\prod_{s=1}^{\infty} \frac{1}{(1-x^s)} = \prod_{s=1}^{\infty} (1+x^s+x^{2s}+\cdots) = \prod_{s=1}^{\infty} \sum_{p_s=0}^{\infty} x^{sp_s} = \sum_{\{p_s=0\}}^{\infty} x^{\sum_{s=1}^{\infty} sp_s} .$$

Thus the coefficient of x^n is the number of ways we can write $n = \sum_{s=1}^{\infty} sp_s$ and this in turn is the number of partitions into p_1 ones, p_2 twos,... .

We can apply the same reasoning to determine the number of partitions $\Pi^{n(m)}$ of n into non-empty boxes of size $\leq m$. Clearly $\sum_{n=0}^{\infty} \Pi^{n(m)} x^n = \prod_{s=1}^{m} (1+x^s+x^{2s}+\cdots)$ or

$$\sum_{n=0}^{\infty} \Pi^{n(m)} x^n = \prod_{s=1}^{m} \frac{1}{1-x^s} .$$

However, there is also a relationship between $\Pi^{n(m)}$ and Π_p^n.

This is most directly obtained by looking at the Ferrers graph of a partition. Consider, e.g., the partition $10 = 5 + 2 + 2 + 1$, a member of Π_4^{10}. We represent this by: the 10 elements being distributed into 4 rows.

The conjugate graph is obtained by interchanging rows and columns. Now the 10 elements are distributed so that the largest box has precisely 4 elements. Graphs of this type do not exhaust

$\Pi^{10(4)}$, which can have partitions with maximum box size less than 4 as well. These would come from original Ferrers graphs of less than 4 rows. Hence we have proven that $\Pi^{10(4)} = \Pi_4^{10} + \Pi_3^{10} + \Pi_2^{10} + \Pi_1^{10} = \Pi_4^{10+4}$, using our first re-recursion, or more generally

$$\Pi^{n(m)} = \Pi_m^{n+m} .$$

The final step of applying the recursion relation can be avoided by subtracting the first row of m dots from the conjugate graph. It is then a partition of $n - m$ into parts of size at most m. Thus $\Pi^{n-m(m)} = \Pi_m^n$ as before.

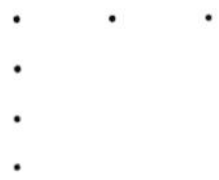

Since a simple graphical proof of the numerous identities among partition numbers of different types is not usually available, analytic means must be found. For example, we have shown above that $\Pi_p^{n+p} = \Pi^{n(p)}$ or

$$\text{coef of } x^n \text{ in } \prod_{s=1}^{p} \frac{1}{1-x^s} = \text{coef of } x^n y^p \text{ in } \prod_{s=0}^{\infty} \frac{1}{1-x^s y} .$$

This is equivalent to the statement that

$$\prod_{s=1}^{p} \frac{1}{1-x^s} = \text{coef of } y^p \text{ in } \prod_{s=0}^{\infty} \frac{1}{1-x^s y},$$

and therefore to the identity

$$\prod_{s=0}^{\infty} \frac{1}{1-x^s y} = \sum_{p=0}^{\infty} y^p \prod_{s=1}^{p} \frac{1}{1-x^s}.$$

The LHS is the function $f(x,y)$ which we have seen satisfies the functional equation and boundary condition

$$(1-y)f(x,y) = f(x,xy)$$
$$f(x,0) = 1.$$

The identity would be established if the RHS which we denote by $\bar{f}(x,y)$, obviously satisfying the boundary condition, also satisfies the functional equation. We have

$$(1-y)\bar{f}(x,y) = \sum_{p=0}^{\infty} \frac{y^p - y^{p+1}}{(1-x)\cdots(1-x^p)} =$$

$$= \sum_{p=0}^{\infty} \left\{ \frac{y^p}{(1-x)\cdots(1-x^p)} - \frac{y^p}{(1-x)\cdots(1-x^{p-1})} \frac{1-x^p}{1-x^p} \right\}$$

$$= \sum_{p=0}^{\infty} \frac{y^p x^p}{(1-x)\cdots(1-x^p)} = \bar{f}(x,xy). \qquad \text{Q.E.D.}$$

Innumerable relations between partition numbers arise from similar identities between generating functions. We list a number of generating functions which can easily be verified.

i) $\dfrac{1}{(1-x)(1-x^3)(1-x^5)\cdots}$ enumerates the partitions into odd parts

ii) $\dfrac{1}{(1-x^2)(1-x^4)(1-x^6)\cdots}$ enumerates the partitions into even parts

iii) $(1+x)(1+x^2)(1+x^3)\ \cdots$ enumerates the partitions into unequal parts

iv) $\dfrac{1}{(1-x)(1-x^4)(1-x^6)(1-x^9)\cdots}$ enumerates the partitions into parts of size $5m \pm 1$.

v) $\dfrac{x^N}{(1-x)(1-x^2)\ \cdots\ (1-x^m)}$ enumerates the partitions of $n - N$ into parts of size $\leq m$.

vi) $\sum_{m=0}^{\infty} \dfrac{x^{m^2}}{(1-x)(1-x^2)\ \cdots\ (1-x^m)}$ enumerates the partitions of n such that successive parts differ in size by at least 2.

Note that iii) follows from $\prod_{s=1}^{\infty}(1+x^s) = \sum_{\{p_s=0\}} x^{\sum_{s=1}^{\infty} sp_s}$. To prove vi) we partition $n - m^2$, using v), into parts of size $\leq m$. We then take the conjugate diagram of at most m rows in non-ascending order and add $2m - 1$ to the first row, $2m - 3$ to the second, ..., 1 to the mth for a total of $m^2 (= 1 + 3 + \cdots + 2m - 1)$. We thus have a partition of n in which the 1st row exceeds the 2nd in size by at least 2, the 2nd similarly with the 3rd, etc.

There are two identities among these 6 generating functions. The first is that i) = iii), because

$$(1+x)(1+x^2)(1+x^3)\ \cdots = \frac{1-x^2}{1-x}\,\frac{1-x^4}{1-x^2}\,\frac{1-x^6}{1-x^3}\cdots$$
$$= \frac{1}{(1-x)}\,\frac{1}{(1-x^3)}\,\frac{1}{(1-x^5)}$$

(the numerator cancels all of the even power terms from the denominator). The second is that iv) = vi), one of the famous Rogers-Ramanujan identities, which is considerably more difficult to prove.

3§. Ramsey's Theorem

Let us interrupt the continuity in order to study a very general structural aspect of partitions, which is also an introduction to partitions on other than a line or set. This is Ramsey's theorem, which tells us how condensed a few component partitions of many objects must be. The most rudimentary form of Ramsey's theorem is known as the pigeonhole principle: If a set S of N elements is partitioned into t components, then at least one component has $\geq [\frac{N-1}{t}] + 1$ elements. Putting it otherwise, for a t component partition of a set of N elements and a given integer q, at least one component will contain a q-element subset of S if $N \geq N_1(q|t) = t(q-1) + 1$.

Ramsey's Theorem

Consider a set S of N elements. Let S_r be the collection of all r-element subsets of S. Partition S_r into $A_1, A_2, \ldots, A_t$. Then given an integer $q \geq r$, there is an integer $N_r(q|t)$ such that if $N \geq N_r(q|t)$, some q-element subset of S has all of its r-subsets in one of the A_j.

Example 1. Consider $r = 1$; this is the pigeonhole principle.

Example 2. Consider $r = 2$. The 2-element subsets of N elements may be represented by lines joining vertices which stand for the elements. A convenient way of partitioning the lines is by coloring each one with one out of t colors. Suppose e.g., $t = 2$ and we represent red by a solid line, blue by a dotted line. Ramsey's theorem would say that if N is large enough some set of q vertices has all of its connecting lines of the same color. If $q = 3$ the counter-example shown proves that $N_2(3|2) > 5$ i.e., there is no triangle of a single color. In fact, it can be shown that $N_2(3|2) = 6$, a fact which is

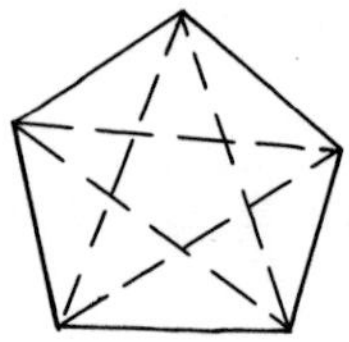

illustrated but not proved by the accompanying diagram.

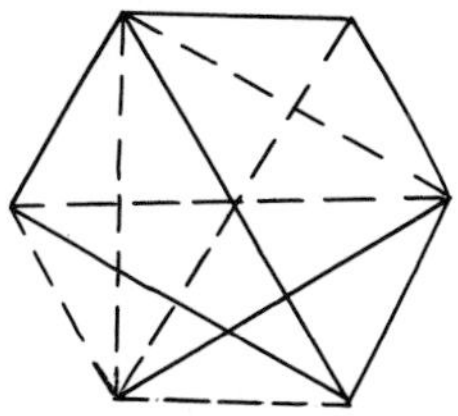

The proof of Ramsey's theorem proceeds most easily in two stages. First, we show that the $t = 2$ case implies the theorem for $t > 2$. In particular, we show that

$$N_r(q|4) \leq N_r(q'|2) \quad \text{where} \quad q' = N_r(q|2).$$

The proof is indicated in the pair of diagrams, in which the notation $S^{[N]}$ signifies that the set S has N elements. Suppose

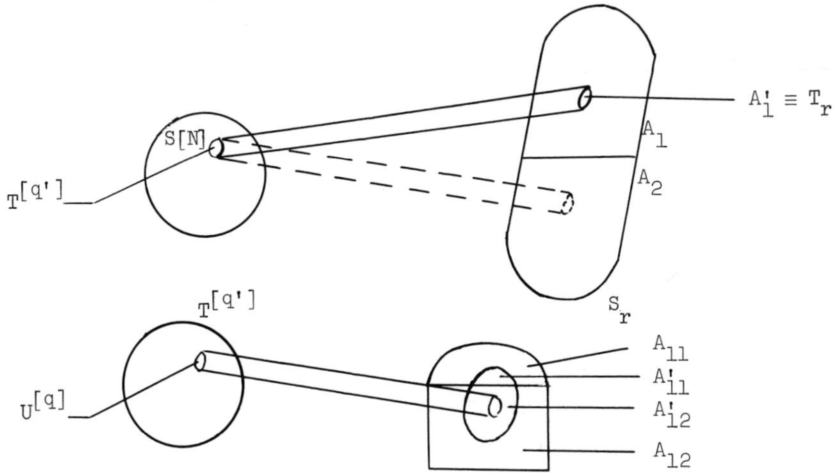

then that $N \geq N_r(q'|2)$. If Ramsey's theorem holds for $t = 2$ then (first diagram) some q'-element set T has all of its r-element subsets T_r contained in either A_1 or A_2 for any division A_1, A_2. Assume without loss of generality that $T_r \equiv A_1' \subset A_1$. Then (second diagram) any subdivision of A_1 into A_{11} and A_{12} will likewise divide T_r into $A_{11}' \subset A_{11}$ and $A_{12}' \subset A_{12}$. But T has $q' = N_r(q|2)$ elements. Hence by Ramsey's theorem for $t = 2$ there is a q-element subset

$U \subset T \subset S$ such that U_r is contained in either A'_{11} or A'_{12}. Thus the fourfold decomposition $S_r = A_{11} \cup A_{12} \cup A_{21} \cup A_{22}$ satisfies Ramsey's theorem for the above value of N.

If we proceed in precisely the same way, we can show that $N_r(q|2^p)$ is bounded by the pth iterate of the operation $q' = N_r(q|2)$. Obviously, $N_r(q|t') \leq N_r(q|t)$ if $t' \leq t$. Hence $N_r(q|t')$ exists for all t' (just choose $2^p > t'$).

The second stage must then consist of proving Ramsey's theorem for $t = 2$. For this purpose, it is convenient to introduce a trivial generalization: Given $r \leq q_1$ and q_2, and a decomposition $S_r = A_1 \cup A_2$ of an N element set S, then if $N \geq N_r(q_1,q_2)$, either some q_1 element subset of S has all r element subsets in A_1 or some q_2 element subset has its r element subsets in A_2. Clearly $N_r(q_1,q_2) \leq N_r(\max q_i|2)$ so that the generalization is not inherently stronger than the original statement.

The proof proceeds by induction. We will, in fact, prove that

$$N_r(q_1,q_2) \leq N_{r-1}(p_1,p_2) + 1$$

where

$$p_1 = N_r(q_1-1,q_2), \qquad p_2 = N_r(q_1,q_2-1).$$

The relevant diagram is shown here. We assume that

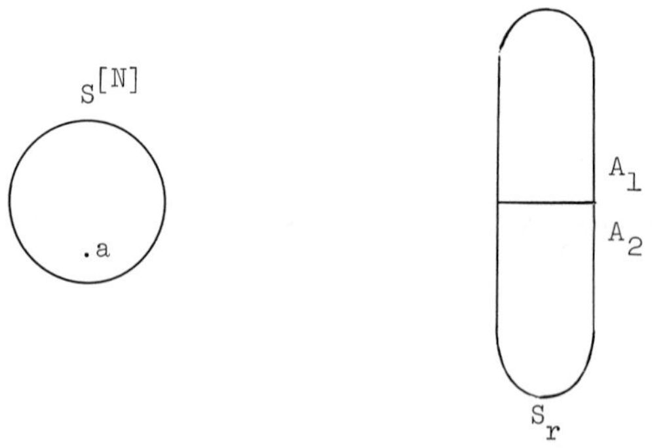

$N \geq N_{r-1}(p_1,p_2) + 1$ and we isolate some fixed element $a \in S$ for the purpose of the induction argument. Now let us excise the element a from S and from those subsets of A_1 and A_2 which contain a. Thus $S = T \cup a$, $x \in B_1 \Longrightarrow x \cup a \in A_1$

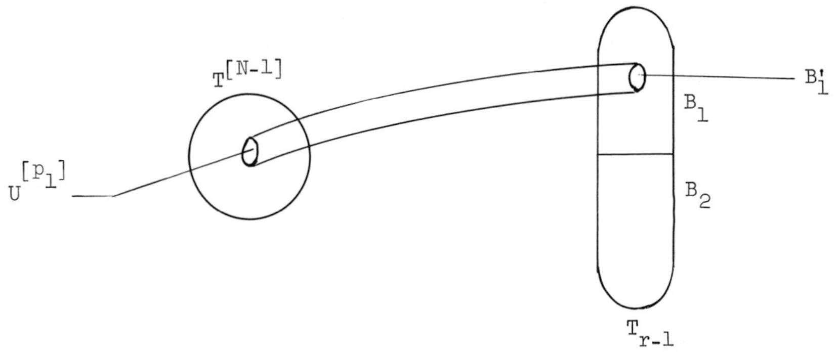

and $x \in B_2 \Longrightarrow x \cup a \in A_2$. Since $N - 1 \geq N_{r-1}(p_1,p_2)$ there is either $U^{[p_1]}$ with its $r - 1$ subsets in B_1 or $U^{[p_2]}$ with its $r - 1$ subsets in B_2. Without loss of generality, suppose the former, so that $U_{r-1} = B_1' \subset B_1$.

Now let us consider the r subsets of U. We suppose that the decomposition of S_r into $A_1 \cup A_2$ divides U_r into $A_1' \cup A_2'$. Since $p_1 = N_r(q_1-1,q_2)$, there are two possibilities:

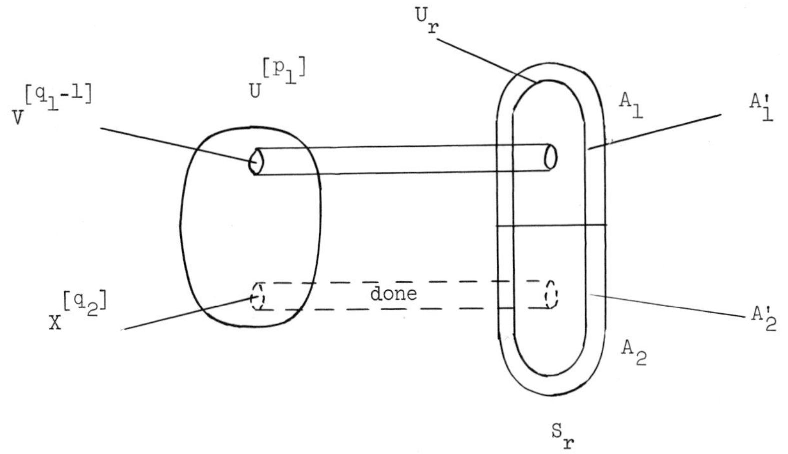

First, there is an $X^{[q_2]}$ such that $X_r \subset A_2' \subset A_2$; then we are done. Second, there is a $V^{[q_1-1]}$ such that $V_r \subset A_1' \subset A_1$. To show that the theorem is satisfied here as well, we must consider two subcases. Define $W = V \cup a$, a q_1-set of S, and let w be an r-subset of W, $w \in W_r$. Then either

i) $a \notin w$. Hence $w \in V_r \subset A_1' \subset A_1$ or

ii) $a \in w$. Hence $w = a \cup v$ where $v \in V_{r-1} \subset W_{r-1} \subset B_1' \subset B_1$, so that $a \cup v \subset A_1$.

The only remaining problem is to show that

$$N_r(q_1,q_2) \leq N_{r-1}(p_1,p_2) + 1 \qquad \text{where} \qquad p_1 = N_r(q_1-1,q_2),\ p_2 = N_r(q_1,q_2-1)$$

is a valid induction process. Let us specify the arguments r, q_1, q_2, by the pair (r, q_1+q_2) in lexicographic order. Since the inequality above either reduces r by 1, or leaves r unchanged and reduces $q_1 + q_2$ by 1, it is only necessary to prove the theorem for the lower bounds of the lexicographic sequence. These occur either when $r = 1$, which is the pigeonhole principle already proved, or when one of q_1 or q_2 is reduced to r. Suppose the former. Then we must show that $N_r(r,q_2)$ exists for $q_2 \geq r$. In fact, it is easily seen that

$$N_r(r,q_2) = q_2, \qquad q_2 \geq r.$$

There are two possibilities.

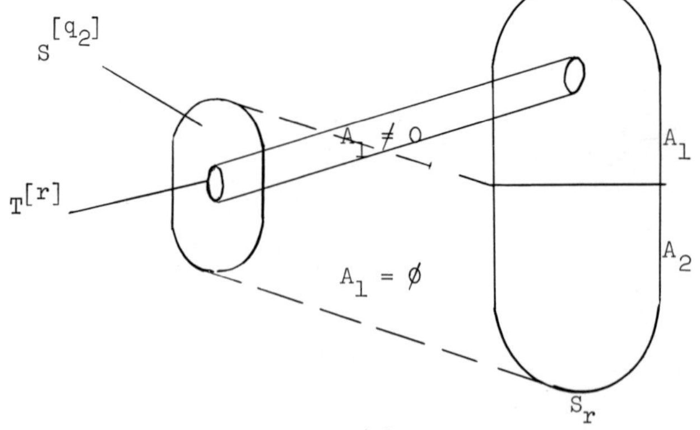

i) $A_1 \neq \emptyset$. Then A_1 contains some r-subset, and therefore all r-subsets of some $T^{[r]} \subset S$.

ii) $A_1 = \emptyset$. Then $A_2 = S_r$, precisely the set of r-subsets of $S^{[q_2]}$. Q.E.D.

Qualitatively, Ramsey's theorem tells us that if r-subsets are given some apriori fixed classification, we cannot stop a large number of sets from falling into one of these categories when the pool of elements becomes very large. Its major use, therefore, is in providing general counter-examples.

Example. A fixed q, every $N \times N$ matrix of 0's and 1's which is large enough, contains a $q \times q$ principle submatrix whose upper triangle is uniform (all 0's or all 1's), as is its lower triangle. To prove this, let S be the set of row vectors $\alpha_i = (a_{ij})$, $j = 1,\ldots,n$. Then consider the pairs of vectors (α_i,α_j), $i < j$, and decompose them into 4 classes depending only on the elements a_{ij} and a_{ji}:

$$(a_{ji}, a_{ij}) = (0,0),\ (1,0),\ (0,1),\ (1,1).$$

According to Ramsey's theorem, if $N \geq N_2(q|4)$, there exists some q-subset of rows such that all pairs of rows are in one class. But this means that all a_{ji} for $i < j$ are either 0 or all 1 and similarly for the a_{ij} with $i < j$.

We conclude by listing in tabular form all known Ramsey numbers $N_2(q_1,q_2)$ [except for the trivial $N_2(2,q_2) = q_2$]. There is also the isolated result $N_2(3,3,3) = 17$.

$q_2 \backslash q_1$	3	4	5
3	6	9	14
4	9	18	
5	14		

4§. Distribution of Labeled Objects

a) Distinguishable boxes

i) Distribute n labeled objects into p distinguishable boxes, with empties permitted. We have only to line up the objects and give each its box number, hence p^n.

ii) Same as i), but empties are not allowed.

Suppose that the sth box contains j_s objects, $s = 1,\ldots,p$. This can be accomplished in $\frac{n!}{j_1!\cdots j_p!}$ different ways, $n!$ for the total number of permutations of the n objects divided by the $j_s!$ indistinguishable arrangements within each box. Thus the required number is given by:

$$c_p^n = \sum_{\substack{\sum_{s=1}^{p} j_s = n \\ \{j_s>0\}_{s=1}^{p}}} \frac{n!}{j_1!\cdots j_p!},$$

expressed most concisely by means of a generating function

$$\sum_{n=0}^{\infty} c_p^n \frac{x^n}{n!} = \sum_{\{j_s>0\}_{s=1}^{p}} \frac{x^{\sum_{s=1}^{p} j_s}}{j_1!\cdots j_p!} = \left(\sum_{j>0} \frac{x^j}{j!}\right)^p = (e^x-1)^p.$$

iii) The total number of arrangements of type ii) of n objects. We define $c^n = \sum_{p=0}^{n} c_p^n$; hence

$$\sum_{n=0}^{\infty} c^n \frac{x^n}{n!} = \sum_{n=0}^{\infty} \sum_{p=0}^{n} c_p^n \frac{x^n}{n!} = \sum_{p=0}^{\infty} (e^x-1)^p = \frac{1}{2-e^x}.$$

b) Collections of pairs -- graph theory [see, e.g., Part B, studies in Statistical Mechanics, Vol. 1, Ed. De Boer and Uhlenbeck, Pub. North Holland].

i) Labeled linear graph: n vertices, labeled $1,\ldots,n$, plus lines connecting certain pairs of vertices.

ii) A linear graph is disconnected if there are two vertex sets which are joined by no line.

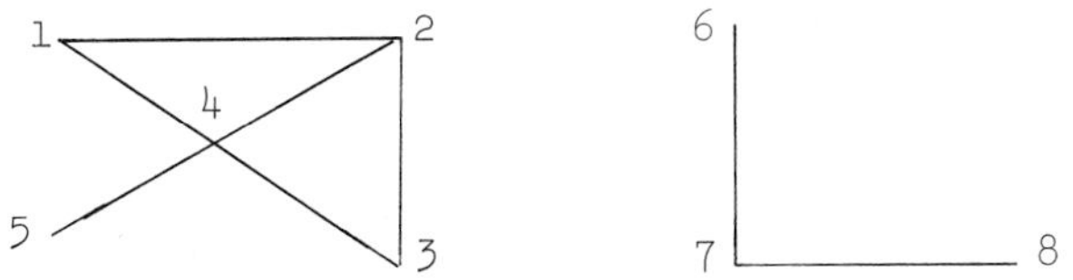

iii) A linear graph which is not disconnected is called connected, e.g., the subgraphs 1 2 3 4 5 and 6 7 8.

iv) A path is a subset of vertices connected successively by lines, e.g., 12345 but not 12435.

v) A cycle is a closed path, i.e., the last vertex is identical with the first, e.g., 1241.

vi) Cycles are independent if no cycle is composed wholly of pieces of the others, e.g., 1241 and 2342 are independent, but 12341 is dependent upon them.

There is then the famous Euler theorem that

$$c = \ell - n + 1$$

where c denotes the number of independent cycles

ℓ " " " " lines

and n " " " " vertices, e.g.

for the graph 12345, we have $2 = 6 - 5 + 1$.

vii) Cayley tree is a connected graph with no cycles, e.g.

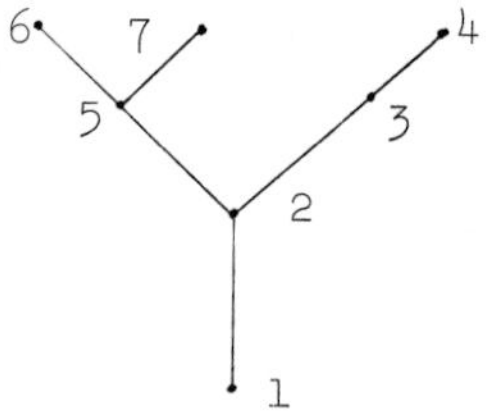

viii) Husimi tree is connected graph in which no line lies on more than one cycle.

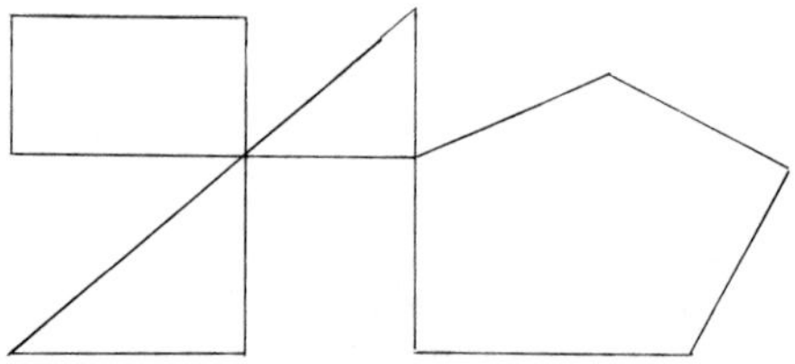

ix) Articulation point is a vertex whose removal, with accompanying lines, disconnects the graph, e.g., vertex 4 in (ii).

x) Star is a graph with no articulation points, e.g.

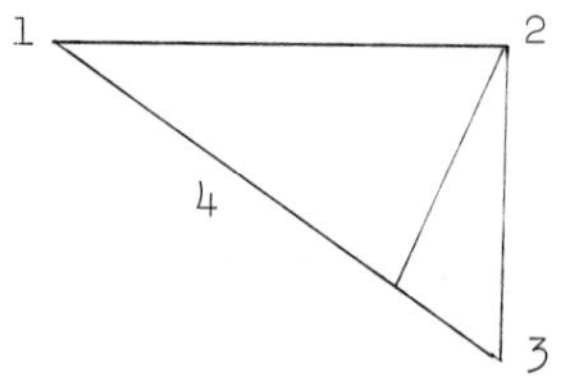

xi) A rooted graph is a graph together with a designated vertex. We will usually indicate this vertex by a circle, e.g. 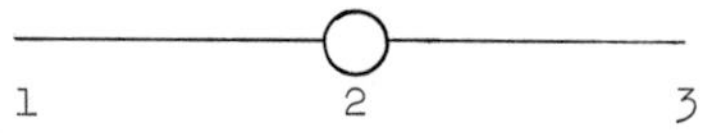

Clearly there are n rooted graphs for each graph on n-vertices.

Let us consider a collection g of linear graphs γ, and associate a weight $W(\gamma)$ with each γ. We shall adopt a sequence of increasingly strong conditions on $W(\gamma)$.

Condition 1. The weight is independent of the labeling. This requires that if γ is a labeled graph in g, so is each permuted graph.

Definition. The counting function for a collection g is

$$g(x) = \sum_n \sum_{\gamma_n \epsilon g} W(\gamma_n) \frac{x^n}{n!}$$

where γ_n refers to a graph of n vertices.

We may also write this as

$$g(x) = \sum_n g_n \frac{x^n}{n!} \quad \text{where} \quad g_n = \sum_{\gamma_n \epsilon g} W(\gamma_n)$$

is the weight of the n vertex subcollection of g.

Definition. The product of a graph on an m-vertex set by a graph on an n-vertex set is the juxtaposition of the two graphs on the combined (m+n)-vertex set with vertices on the second graph now being marked $m + 1,\ldots,m + n$. The product $g * h$ of two collections of graphs g and h is obtained as follows:

i) Each $\gamma_m \in g$ is a member of a set of $m!$ permuted graphs. Select one of these.

ii) Carry out the same operation for γ'_n of h.

iii) Construct the collection of products of one graph from each of i) and ii) together with all $(m+n)!$ permutations. Note that the asymmetry in the definition of the product of two graphs is now irrelevant, since the product of two collections is certainly commutative.

Condition 2. The weight of a product graph is a product of the component weights.

The product theorem: $(g * h)(x) = g(x)h(x)$.

Proof: Since $g_n = \sum_{\gamma_n \epsilon g} W(\gamma_n)$ then $\frac{1}{n!} g_n = \sum_{\{\gamma_n \epsilon g\}} W(\gamma_n)$ where $\{\gamma_n \in g\}$ signifies that only one representative of each permutation of $1,\ldots,n$ is chosen. Hence by the definition of product collection, the total weight of single representatives of $(g * h)_n$ is given by

$$\frac{1}{n!}(g * h)_n = \sum_{n_1+n_2=n} \sum_{\{\gamma_{n_1} \epsilon g, \gamma'_{n_2} \epsilon h\}} W(\gamma_{n_1} * \gamma'_{n_2}) = \sum_{n_1+n_2=n} \frac{1}{n_1!} g_{n_1} \frac{1}{n_2!} h_{n_2} .$$

Therefore,

$$(g * h)(x) = \sum_n \sum_{n_1+n_2=n} x^{n_1+n_2} g_{n_1} h_{n_2} \frac{1}{n_1! n_2!} . \qquad \text{Q.E.D.}$$

A disconnected graph is (to within permutation) the product of its connected components. This suggests

Condition 3. The weight of a disconnected graph is the product of the weights of its connected components.

Theorem. Let f be a collection of connected graphs, and F the collection of graphs whose connected components are chosen from f. Then $e^{f(x)} = 1 + F(x)$.

Proof: Let $F_m(x)$ be the counting function for graphs of exactly m connected components. Then, since an m component graph $\gamma_1 * \gamma_2 * \cdots * \gamma_m$ occurs in $[f(x)]^m$ $m!$ times, we clearly have

$$F_m(x) = \frac{1}{m!}[f(x)]^m.$$

Hence

$$F(x) = \sum_{m=1}^{\infty} F_m(x) = \sum_{m=1}^{\infty} \frac{1}{m!}[f(x)]^m = e^{f(x)} - 1. \qquad \text{Q.E.D.}$$

Example. Find $C(n,k)$, the number of connected graphs of n labeled vertices with k lines.

Let $N(n,k)$ be total number of such graphs, connected or not. Then

$$N(n,k) = \binom{\binom{n}{2}}{k} = \binom{\frac{1}{2}n(n-1)}{k} .$$

If we use a weight of y for each line, then $C(n,k)$ will be given by the coefficient of $\frac{1}{n!}x^n y^k$ in the connected graph counting function. But

$$N(x,y) = \sum_{n=1}^{\infty} \frac{x^n}{n!} \sum_{k=0}^{\frac{1}{2}n(n-1)} \binom{\frac{1}{2}n(n-1)}{k} y^k$$

$$= \sum_{n=1}^{\infty} \frac{x^n}{n!} (1+y)^{\frac{1}{2}n(n-1)} .$$

It follows from the previous theorem that

$$C(x,y) = \log(1 + N(x,y)) = \log\left(1 + \sum_{n=1}^{\infty} \frac{x^n}{n!} (1+y)^{\frac{1}{2}n(n-1)}\right)$$

$$= \sum_{r=1}^{\infty} (-1)^{r-1} \frac{1}{r} \left[\sum_{n=1}^{\infty} \frac{x^n}{n!} (1+y)^{\frac{1}{2}n(n-1)}\right]^r$$

$$= \sum_{r=1}^{\infty} (-1)^{r-1} \frac{1}{r} \sum_{\substack{n_j=1 \\ j=1..r}}^{\infty} \frac{x^{\sum_{j=1}^{r} n_j}}{\prod_{j=1}^{r} n_j!} (1+y)^{\frac{1}{2}\sum_{j=1}^{r} n_j(n_j-1)}$$

$$= \sum_{n,k} C(n,k) \frac{x^n y^k}{n!}$$

Hence

Hence

$$C(n,k) = \sum_{r=1}^{\infty} \frac{(-1)^{r-1}}{r} \sum_{\sum_{j=1}^{r} n_j = n} \frac{n!}{\prod_{j=1}^{r} n_j!} \binom{\frac{1}{2}\sum_{j=1}^{r} n_j(n_j-1)}{k}$$

As $n \to \infty$ the series starts as:

$$C(n,k) = \binom{\frac{1}{2}n(n-1)}{k} - n\binom{\frac{1}{2}(n-1)(n-2)}{k} + \cdots$$

which, if $\frac{k}{n \log n}$ is fixed, is dominated by its first term, precisely $N(n,k)$.

We have seen that when a graph is decomposed into its connected components, there is a natural choice of its weight which extends to a simple relationship between the counting functions of all graphs and those of connected graphs. We now carry this process one step further and relate connected graphs to even smaller constituents.

Definition. Star-tree decomposition of a connected graph: remove all articulation points, separate the disconnected figures which result, and resupply each one with its lost vertices. There results a collection of stars (with total number of vertices $\geq$ that of the original graph) connected together in the fashion of a tree. e.g.

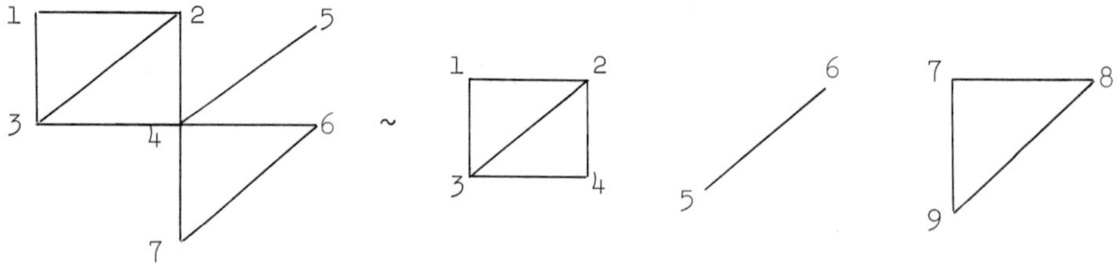

If all of the resultant stars were polygons, the original graph would, in fact, be a

Husimi tree; if all stars were two-vertex lines, it would be a Cayley tree.

Condition 4. The weight of a connected graph is the product of the weights of the stars in its star-tree decomposition.

Theorem. Let S be a collection of stars and T the collection of rooted graphs whose component stars belong to S. Then

$$T(x) = xe^{S'(T(x))}.$$

Proof: Consider a rooted graph. The root is an articulation point of some order m (the number of connected components resulting from its removal). Hence if $M(x)$

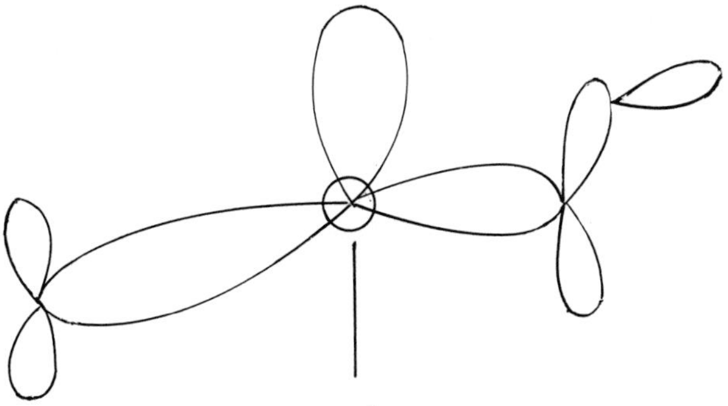

is the counting function for a single branch emanating from the root (omitting the root vertex) we must have

$$T(x) = xe^{M(x)}.$$

Now consider the star-tree decomposition of any graph γ in M, and let S_γ be that star which contains the root. We see that S_γ has one vertex missing -- the root -- while any other vertex is the root of some rooted graph. If S_γ has q vertices and weight function Wx^q, the weight of all branches built from S_γ must then be

$qW[T(x)]^{q-1}$. Hence

$$M(x) = \frac{d}{dT(x)} S(T(x)). \qquad \text{Q.E.D.}$$

It is useful to note that if t denotes the collection of graphs corresponding to the rooted collection T, and if

$$t(x) = \sum_{n=1}^{\infty} \sum_{C_n \in T} W(C_n) \frac{x^n}{n!}$$

then

$$T(x) = \sum_{n=1}^{\infty} \sum_{C_n \in T} W(C_n) \frac{x^n}{(n-1)!} = xt'(x).$$

Example. Find the total number T_n of rooted Cayley trees on n vertices. Here the only component stars are two-vertex lines so that $S(x) = \frac{1}{2!} x^2$ and $S'(x) = x$. The tree counting function hence satisfies

$$T(x) = xe^{T(x)},$$

which may be solved by the Lagrange expansion theorem: If

$$y = a + f(y)$$

then explicitly

$$y = a + \sum_{n=1}^{\infty} \frac{1}{n!} \frac{d^{n-1}}{dy^{n-1}} (f(y))^n \Big|_{y=a}.$$

Since

$$T = 0 + xe^T,$$

we have

$$T = 0 + \sum_{n=1}^{\infty} \frac{1}{n!} \left(\frac{d}{dT}\right)^{n-1} (xe^T)^n \Big|_{T=0}$$

$$= \sum_{n=1}^{\infty} \frac{1}{n!} \left(\frac{d}{dT}\right)^{n-1} (x^n e^{nT}) \Big|_{T=0}$$

$$= \sum_{n=1}^{\infty} \frac{1}{n!} x^n n^{n-1} e^{nT} \Big|_{T=0} = \sum_{n=1}^{\infty} \frac{n^{n-1}}{n!} x^n .$$

We conclude that

$$T_n = n^{n-1}$$

and the total number of unrooted trees t_n is

$$t_n = \frac{T_n}{n} = n^{n-2} .$$

c) <u>Indistinguishable boxes (and labeled objects)</u>.

Such distribution problems are solved in precisely the same way as partition problems, but the resulting generating functions are uniformly simpler in appearance. For example: Find the number of ways V_n of placing n labeled objects into non-empty indistinguishable boxes. Suppose that n_j boxes have occupation j (contain j objects apiece). Then $n = n_1 \cdot 1 + n_2 \cdot 2 + n_3 \cdot 3 + \cdots$ and the number of ways of partitioning the n labeled objects in this fashion is

$$V(\{n_j\}) = \frac{n!}{(1!)^{n_1}(2!)^{n_2}(3!)^{n_3} \cdots n_1! n_2! n_3! \cdots} ,$$

since permutation of boxes with the same number of objects, or or all objects within a box, does not change the state of the system. Clearly

$$\sum_{n=0}^{\infty} V_n \frac{x^n}{n!} = \sum_{n=0}^{\infty} \frac{x^n}{n!} \sum_{\sum_{j=1}^{\infty} jn_j = n} \frac{n!}{\prod_{j=1}^{\infty} (j!)^{n_j} \prod_{j=1}^{\infty} n_j!}$$

$$= \sum_{n=0}^{\infty} \sum_{\sum_{j=1}^{\infty} jn_j = n} \frac{x^{\sum_{j=1}^{\infty} jn_j}}{\prod_{j=1}^{\infty} (j!)^{n_j} \prod_{j=1}^{\infty} n_j!} = \prod_{j=1}^{\infty} \sum_{n_j=0}^{\infty} \frac{x^{jn_j}}{(j!)^{n_j} (n_j)!}$$

$$= \prod_{j=1}^{\infty} e^{\frac{x^j}{j!}} = e^{\sum_{j=1}^{\infty} \frac{x^j}{j!}} = e^{(e^x - 1)} ,$$

the desired generating function.

d) Partially labeled graphs -- The Polya Theorem

Consider the 5 line diagram shown

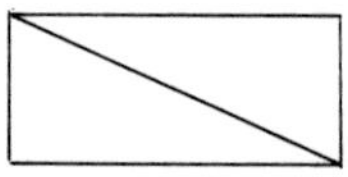

Since it has 4 vertices there are 4! ways of distributing vertex numbers 1-4. However, it is up to us to specify which of the 24 are to be regarded as distinct. If diagrams which transform to each other under a plane rotation are equivalent, we have only 12 distinct diagrams, e.g.,

are equivalent. The number of labeled graphs, on the other hand, is only 6. A labeled graph we recall is specified by those lines which connect vertices 1-4, an explicit representation being given by the symmetric connection matrix, i.e. for

the graph 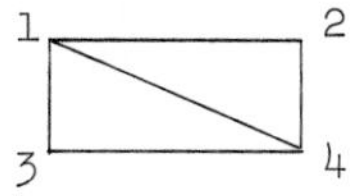 ,

$$\begin{array}{c} \\ 1 \\ 2 \\ 3 \\ 4 \end{array} \begin{array}{c} \begin{array}{cccc} 1 & 2 & 3 & 4 \end{array} \\ \begin{pmatrix} \cdot & 1 & 1 & 1 \\ 1 & \cdot & 0 & 1 \\ 1 & 0 & \cdot & 1 \\ 1 & 1 & 1 & \cdot \end{pmatrix} \end{array}$$

By permuting vertices all we can do is place the single zero (the absence of a connecting line) in one of the 6 available locations. The diagrams which are inequivalent under spacial rotations also number 6, it is clear. Finally, if any two graphs which differ only by a permutation of vertices are regarded as equivalent, then our original diagram yields only one distinct graph.

There are many situations in counting of graphs when graphs which are equivalent under some prescribed groups of operations are to be counted only as a single distinct graph. The relation between the total number of labeled graphs and the number of distinct graphs depends intimately upon the structure of the equivalence group, and may be obtained by applying a basic theorem due to Polya. This theorem deals with a collection f of figures $\{\phi\}$, any one of which may be attached (with repetition) to any of p given vertices. The p vertices with figures attached to each will be referred to as a configuration $\overline{\Phi}$. We are also given a permutation group G on the p vertices, and we define two configurations as equivalent if some permutation, in G, of the p vertices with their attached figures converts one configuration into the other. The problem is to count the inequivalent configurations.

We start by counting the figures: define

$$f(\underset{\sim}{x}) = \sum_{\phi \epsilon f} W_{\underset{\sim}{x}}(\phi),$$

where $W_{\underset{\sim}{x}}(\phi)$ is a weight function which depends upon a parameter (or set of parameters) $\underset{\sim}{x}$. We further define the weight of a configuration as the product of the weights of its component figures, and construct the corresponding counting function

$$F(\underset{\sim}{x}) = \sum_{\underline{\phi} \in F} W_{\underset{\sim}{x}}(\underline{\phi}),$$

where F contains the inequivalent configurations constructed from f. Finally, we introduce the cycle index of G, a p-variable function defined by

$$Z(G|z_1,\ldots,z_p) = \frac{1}{g} \sum_{j_1,\ldots,j_p} g_{j_1,\ldots,j_p} (z_1)^{j_1} \cdots (z_p)^{j_p},$$

where g is the order of the group G and $g_{j_1,\ldots,j_p}$ is the number of permutations in G which consist of j_1 cycles of length 1, j_2 length $2,\ldots,$ j_p of length p. $(1 \cdot j_1 + 2 \cdot j_2 + \cdots + p \cdot j_p = p)$. Then the Polya theorem states that

$$F(\underset{\sim}{x}) = Z(G|f_1(\underset{\sim}{x}),\ f_2(\underset{\sim}{x}),\ldots,f_p(\underset{\sim}{x})),$$

where $f_j(\underset{\sim}{x}) = \sum_{\phi \in f} [W_{\underset{\sim}{x}}(\phi)]^j$.

As an example of the explicit form of the cycle index, consider S_3, the full permutation group on 3 letters. Since the permutations, in cycle notation, are given by

$$e = (1)(2)(3),\quad (12)(3),\quad (13)(2),\quad (23)(1),\quad (123),\quad (132),$$

we have

$$Z(S_3|z_1,z_2,z_3) = \frac{1}{3!}\,(z_1^3+3z_1z_2+2z_3).$$

Examples.

i) Our first example of the Polya theorem itself relates to a simple but very artificial situation. We consider the 6 vertices of a regular octahedron (e.g., the 6 points on the 3 coordinate axes which are one unit from the origin) and ask for the number of ways of placing 3 red balls, 2 blue balls, and 1 yellow ball on the 6 vertices. Any rotation which carries the octahedron into itself defines an equivalent configuration.

To solve this problem we regard the colors that a ball, at a given vertex, may take as the collection of figures. If a red ball is given the weight x, a blue y, and a yellow z, then the counting function is, clearly,

$$f(x,y,z) = x + y + z.$$

The required number of configurations is then the coefficient of x^3y^2z in $F(x,y,z)$. It is readily verified that the cycle index for the octahedral group is

$$Z = \frac{1}{24}\left(z_1^6 + 6z_1^2z_4 + 3x_1^2z_2^2 + 6z_2^3 + 8z_3^2\right).$$

Substituting

$$z_1 = f_1(x,y,z) = x + y + z$$
$$z_2 = f_2(x,y,z) = x^2 + y^2 + z^2$$
$$z_3 = f_3(x,y,z) = x^3 + y^3 + z^3,$$

only the first and third terms of the cycle index contribute to x^3y^2z, and so we have

$$\# = \text{coefficient of } x^3y^2z \text{ in}$$

$$F(x,y,z) = \frac{1}{24}\left((x+y+z)^6 + 3(x+y+z)^2(x^2+y^2+z^2)^2 + \cdots\right) = 3.$$

ii) As a more general example suppose that all vertices are distinguishable, so that the symmetry group consists of the identity $e = (1)^p$ alone. Then according to Polya's theorem,

$$F(\underset{\sim}{x}) = (f(\underset{\sim}{x}))^p,$$

a result which was previously obtained.

iii) At the other extreme, suppose that all vertices are equivalent, so that the permutation group is the full symmetric group S_p. We must now count all permutations. In cycle notation, each permutation is a partition of $1 \cdots p$ into non-empty boxes. Consider a structure consisting of j_1 cycles of length 1, j_2 cycles of length 2,..., j_p cycles of length p. The p! permutations of this structure will cause $j_1!j_2! \ldots j_p!$ repetitions by interchanging cycles of the same length and $1^{j_1}2^{j_2} \ldots p^{j_p}$ repetitions because there are k equivalent ways of representing a cycle of length k, i.e., (1234) = (2341) = (3412) = (4123). Hence

$$Z(S_p|z_1,\ldots,z_p) = \frac{1}{p!} \sum_{\Sigma\, kj_k=p} \frac{p!}{j_1!\ldots j_p!\, 1^{j_1}\ldots p^{j_p}} z_1^{j_1} \ldots z_p^{j_p},$$

and we conclude that

$$F(\underset{\sim}{x}) = \sum_{\Sigma\, kj_k=p} \prod_{k=1}^{p} \left[\frac{1}{j_k!} \left(\frac{f_k(\underset{\sim}{x})}{k}\right)^{j_k}\right].$$

Proof of Polya's Theorem

We relate the sum over inequivalent configurations to a sum over all configurations which satisfy some part of the symmetry of the group. Consider a fixed configuration Φ. We divide the group G into equivalence classes according to their action upon Φ: $P \sim Q \Longleftrightarrow P\Phi = Q\Phi$. If $Q_1 = e, Q_2,\ldots,Q_{h_\Phi}$ are a complete set of inequivalent elements of G and $G_\Phi = \{P | P\Phi = \Phi\}$, then $G = \bigcup_{\alpha=1}^{h_\Phi} (Q_\alpha G_\Phi)$. Correspondingly, since $Q_\alpha G_\Phi \Phi$ consists of g_Φ replicas of the configuration $Q_\alpha \Phi$, we have $\{P\Phi\} = g_\Phi \{Q_\alpha \Phi\}$: each of the $h_\Phi = \frac{g}{g_\Phi}$ distinct configurations equivalent to Φ occurs g_Φ times.

Now we consider $\sum_{P\in G} \sum_{\{\Phi | P\Phi=\Phi\}} W(\Phi)$ and ask how often a distinct configuration Φ occurs in this sum. Since a given distinct Φ satisfies $P\Phi = \Phi$ for precisely g_Φ P's, the second sum $\left(\sum_{\{\Phi | P\Phi=\Phi\}} W(\Phi) \right)$ occurs g_Φ times. But this second sum contains not just each Φ, but $\frac{g}{g_\Phi}$ distinct configurations equivalent to it. We conclude that a distinct configuration occurs precisely g times in the double sum, so that $F = \frac{1}{g} \sum_{P\in G} \sum_{\{\Phi | P\Phi=\Phi\}} W(\Phi)$.

Let us evaluate

$$F_p = \sum_{\{\Phi | P\Phi=\Phi\}} W(\Phi).$$

Suppose that the permutation P is decomposed into cycles $P = \prod_s c_s$. To satisfy $P\Phi = \Phi$, the vertices of P corresponding to each cycle must have identical figures attached. Hence

$$F_p = \prod_s \left[\sum_\phi W(\phi)^{d(c_s)} \right],$$

where $d(c_s)$ is the length of the cycle c_s, and so

$$F = \frac{1}{g} \sum_{\prod_s c_s \in G} \prod_s \left[\sum_\phi W(\phi)^{d(c_s)} \right]$$

$$= \frac{1}{g} \sum_{j_1, j_2, \ldots j_p} g_{j_1 j_2 \cdots j_p} \left(\sum_\phi W(\phi) \right)^{j_1} \cdots \left(\sum_\phi W(\phi)^p \right)^{j_p}. \qquad \text{Q.E.D.}$$

iv) Find the number R_n of isomers of saturated alcohols $C_nH_{2n+1}OH$. This is the number of rooted Cayley trees in which the degree of the root is ≤ 3 and the degree of all other vertices is ≤ 4 [Degree of vertex $\equiv$ number of lines joining at the vertex.]

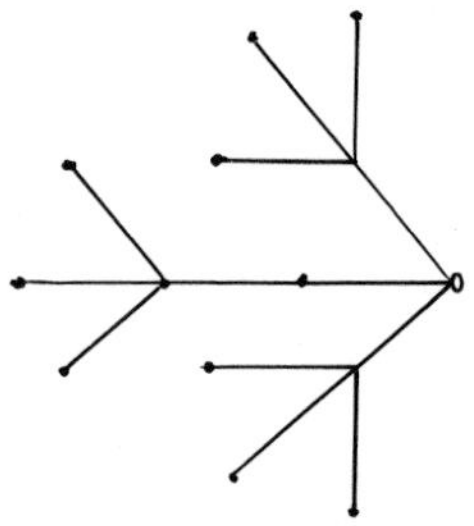

The chemical structure corresponding to a given tree places hydrogens at each vertex for a total valence of 4, except for the root which must have one hydroxyl group:

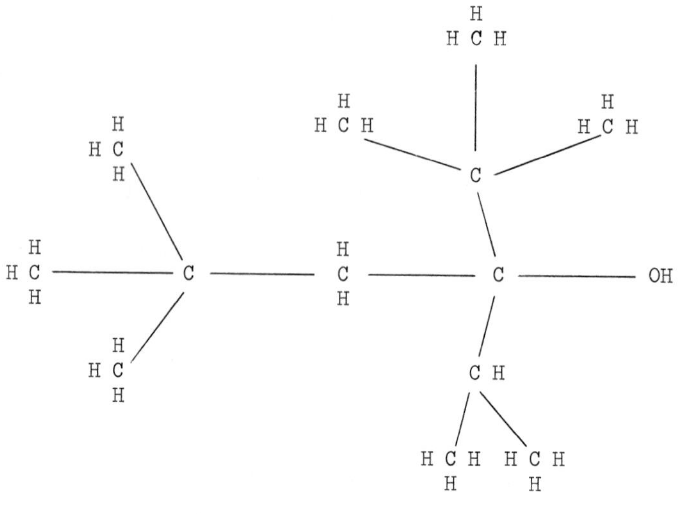

To solve this problem we weight each carbon vertex by x and set $R_0 = 1$, with $R(x) = \sum_{n=0}^{\infty} R_n x^n$. We then build $R(x)$ by hanging, from the 3 available vertices adjoining the root, 3 smaller rooted graphs of the same type. (I.e., alcohols without the OH group.)

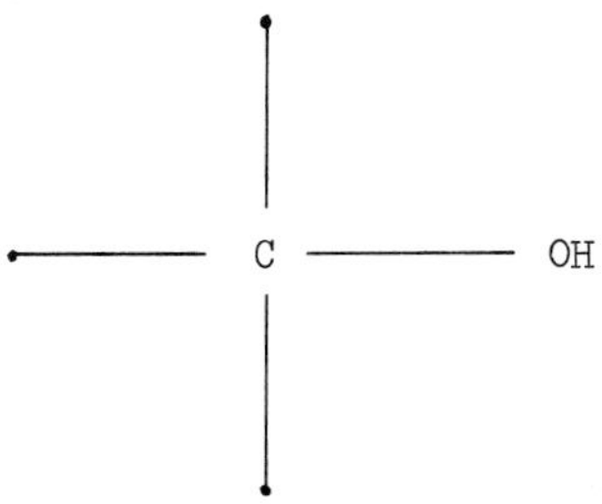

Since the order of the side chains is immaterial, the symmetry group here is the full symmetric group on 3 letters and the figures are themselves saturated alcohols. Clearly,

$$R_1(x) = \sum_{n=0}^{\infty} R_n x^n = R(x)$$

$$R_2(x) = \sum_{n=0}^{\infty} R_n (x^2)^n = R(x^2)$$

$$R_3(x) = \sum_{n=0}^{\infty} R_n (x^3)^n = R(x^3),$$

and we, therefore, obtain from the Polya theorem the recursion relation

$$R(x) = 1 + xZ(S_3 | R_1(x), R_2(x), R_3(x) = 1 + \frac{x}{6} [(R(x))^3 + 3R(x)R(x^2) + 2R(x^3)].$$

Note that the weight x in front of Z is that of the root which is not counted in the sum over configurations.

v) Find the number D_n of stereo isomers of the saturated alcohols. This is a counting problem in which the orientation of the graph is important: only cyclic permutations of the 3 branches of the root define equivalent structures. In precisely the fashion of the preceding example:

$$D(x) = 1 + xZ(c_3 | D_1(x),\ D_2(x),\ D_3(x)),$$

where C_3 is the cyclic group on 3 letters. Since the permutations of the cyclic group are e, (123), (132), the cycle structure is $\{(1)^3, 2(3)\}$ and so

$$D(x) = 1 + \frac{x}{3}((D(x))^3 + 2D(x^3)).$$

vi) Find the number T_n of n-line rooted Cayley trees with no degree restriction on the vertices.

Let $T^{(m)}$ be the number of n-line Cayley trees with m branches from the root, with $T_0^{(m)} = \delta_{m,0} = T_m^{(0)}$. We weight each line by x and define

$$T^{(m)}(x) = \sum_{n=0}^{\infty} T_n^{(m)} x^n.$$

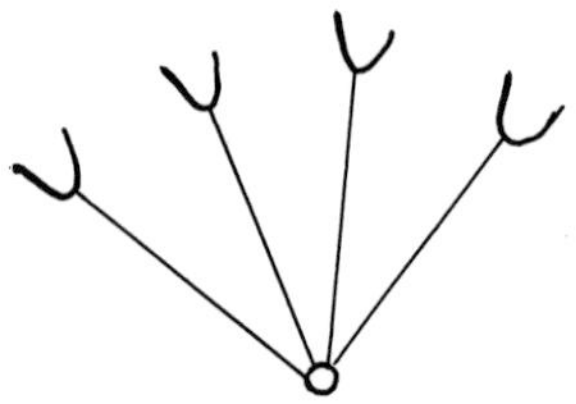

Clearly,

$$T^{(m)}(x) = x^m Z(S_m | T(x), T(x^2), \ldots, T(x^m))$$

whence from Example iii)

$$T(x) = \sum_{m=0}^{\infty} T^{(m)}(x) = \sum_{m=0}^{\infty} x^m \sum_{\Sigma kj_k = m} \prod_{k=1}^{\infty} \frac{1}{j_k!} \left(\frac{T(x^k)}{k}\right)^{j_k},$$

where the terms for which $k > m$ do not contribute anyway. We get

$$T(x) = \sum_{j_1,\ldots,j_k\ldots=0}^{\infty} \frac{x^{\sum_{k=1}^{\infty} kj_k} \prod_{k=1}^{\infty} [T(x^k)]^{j_k}}{\prod_{k=1}^{\infty} j_k! \prod_{k=1}^{\infty} k^{j_k}},$$

but

$$\sum_{j=0}^{\infty} \frac{x^{kj}(T(x^k))^j}{j!k^j} = \exp \frac{x^k T(x^k)}{k},$$

so that

$$T(x) = \exp \sum_{k=1}^{\infty} \frac{x^k T(x^k)}{k},$$

which is Cayley's functional equation.

Exercises.

1) Show that c_p^n, the number of distributions of n labeled objects into p non-empty labeled boxes, can be written as

$$c_p^n = \Delta^p 0^n \equiv \sum_0^p (-1)^k \binom{p}{k} (p-k)^n.$$

2) From the relation

$$c_{m,n} + mc_{m-1,n} + \cdots + \binom{m}{k} c_{m-k,n} + \cdots + c_{0,n} = \binom{n+m-1}{n}$$

for distributions of n indistinguishable objects into m non-empty labeled boxes, show that

$$c_{m,n} = \sum_{k=0} (-1)^k \binom{m}{k} \binom{n+m-k-1}{n}.$$

3) From the identity

$$(1-x)^{-1} = (1+x)(1+x^2)(1+x^4) \cdots (1+x^{2n}) \cdots,$$

show that every number as a unique expression in the binary system. From the same identity, show that every number greater than 1 has the same number of partitions into an even number of parts of the form $1,2,4,\ldots,2^n,\ldots$ as into an odd number of parts.

4) Employing the generating function for non-empty compositions of n with no class exceeding s elements, show that

(a) the $c_n^{(2)}$ are the Fibonacci numbers

(b) $c_n^{(s)} - 2c_{n-1}^{(s)} + c_{n-1-s}^{(s)} = \delta_{1,n} - \delta_{s+1,n}$.

5) Find the total number of unlabeled Cayley trees with n vertices.

6) Find the number of 2-terminal series-parallel n-vertex graphs.

(Terminal = root, at most one line joins 2 vertices, every component of a spindle is a chain and every component of a chain is a spindle:

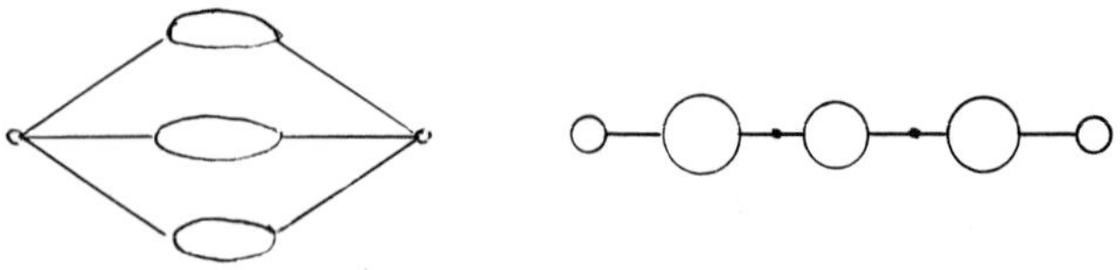

).

spindle chain

e) Counting unrooted (free) unlabeled graphs

We recall that to each labeled n-vertex graph, there correspond precisely n rooted labeled n-vertex graphs. Because of equivalence, there is no such obvious relationship for unlabeled graphs, e.g.,  gives rise only to 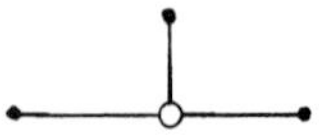 and 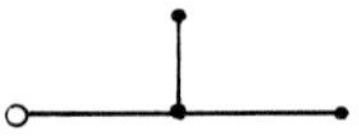. Such graphs can be counted by building them up from an artificial central root, or more easily by use of the Dissimilarity theorem. The theorem deals with pure star-trees, i.e., star-trees with all stars the same. Consider the group G of vertex permutations which leave a given graph (i.e., its connecting lines) unaltered. Two points which are interchanged by some $P \in G$ will be called <u>similar</u>; two stars interchanged by some $P \in G$ will also be called <u>similar</u>.

Dissimilarity Theorem

Suppose there are p similarity classes of points, represented by p dissimilar points. Also s similarity classes of stars represented by s dissimilar stars. Let p_k be the number of dissimilar points in the kth dissimilar star. Then

$$1 = p - \sum_{k=1}^{s} (p_k - 1).$$

The proof is trivial. First remove an end star and all similar stars, leaving their articulation points. Hence $p_1 - 1$ dissimilar points have been removed and $s - 1$ dissimilar stars remain. Continue the pruning process until only one star is left with p_s dissimilar points. Hence $p - p_s = \sum_{k=1}^{s-1} (p_k - 1)$. Q.E.D.

To make use of this theorem, we first observe that it still holds if the stars are not identical but are chosen from a given star collection. Now consider the unrooted star-trees T_n built out of n stars of a given collection. Any dissimilar point of such a star tree can be chosen as a root. Hence if the theorem is summed over all n-star trees from the collection, with any suitable weight function, it reads

$$\theta_n = \Theta_n - A_n + B_n,$$

where

θ_n is the total weight of trees

Θ_n " " " " " rooted trees

$$A_n = \sum_{\{T_n\}} \sum_{k=1}^{s} p_k W(T_n)$$

$$B_n = \sum_{\{T_n\}} s\, W(T_n).$$

The trick then is that A_n and B_n can usually be obtained from θ_n via Polya's theorem.

Example. Find the number t_n of free Cayley trees of n lines. Here B_n is the number of dissimilar lines in all n-line Cayley trees. It may be obtained by fixing one line and hanging rooted Cayley trees on each end for a total of n lines. Since left-right reflection about the center of the line does not produce a different graph, the equivalence group consists of the identity plus a reflection. Hence

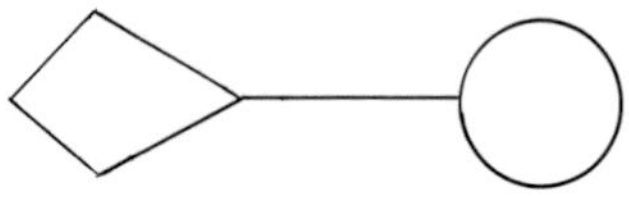

$$B(x) = xZ(S_2 | T(x), T(x^2)) = x(\frac{1}{2}(T^2(x) + T(x^2))).$$

On the other hand, A_n is the number of dissimilar points, weighted by the number of dissimilar lines each one is on. Now a reflection shifts the point from the left end of the line to the right, giving it precisely the correct weight. Hence $A(x) = xT^2(x)$. We conclude that

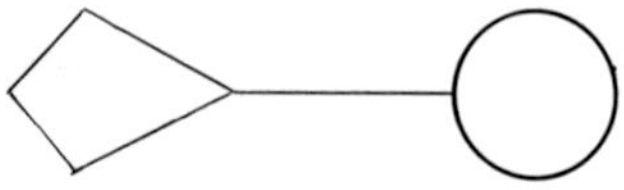

$$t(x) = T(x) - \frac{1}{2}x[T^2(x) - T(x^2)].$$

II. COUNTING AND ENUMERATION OF A REGULAR LATTICE

(Maximum Geometric Structure)

A. Random Walk on Lattices.

A regular lattice is an infinite array of points

$$\underset{\sim}{x} = \ell_1 \underset{\sim}{a}_1 + \ell_2 \underset{\sim}{a}_2 + \ell_3 \underset{\sim}{a}_3 + \underset{\sim}{x}_0 \qquad \text{(3 dim lattice)}$$

where the $\underset{\sim}{a}_i$'s are linearly independent base vectors, the ℓ_i's are integers, and there are as many $\underset{\sim}{x}_0$'s as the number of interpenetrating lattices. For example, in 2-dimensions, we have the square and equilateral triangular lattices:

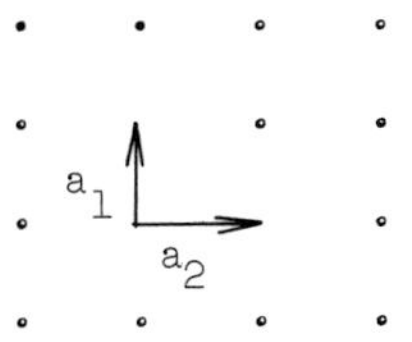

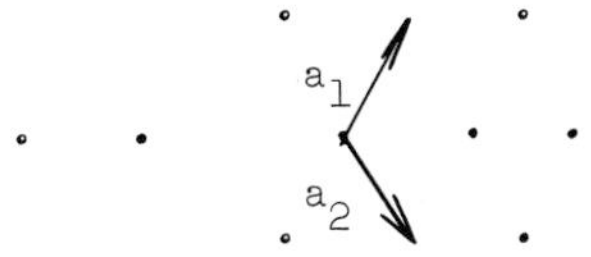

where a single $\underset{\sim}{x}_0$ is required, which may be chosen as the origin. On the other hand, for the regular hexagonal lattice

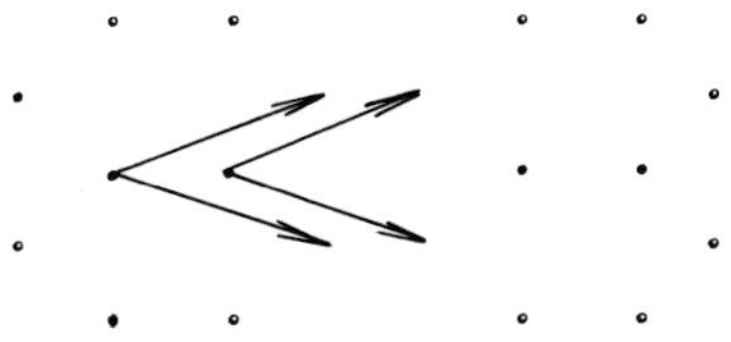

we require two interpenetrating triangular lattices as shown. Even when it is not necessary, it may be convenient to use interpenetrating lattices, e.g., the A-site B-site decomposition of a square lattice:

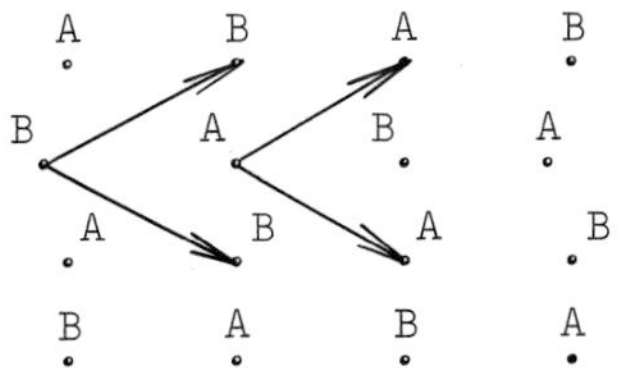

The lattice vertices constitute the space from which points will be chosen, and questions will be asked with particular reference to the order relations which exist in the lattice.

1§. Regular Cubic Lattices.

Consider for the time being only s-dimensional cubic lattices of unit side. By a random walk we mean a path on the lattice in which each step is a unit along a coordinate direction, i.e., each step is from a vertex to a nearest neighbor. If a location $(\ell_1,\ell_2,\ldots,\ell_s)$ is given the generating function designation $z_1^{\ell_1} z_2^{\ell_2} \ldots z_s^{\ell_s}$ then a step may be made to one of the locations:

$$\begin{array}{lll} z_1^{\ell_1 \pm 1} & z_2^{\ell_2} \ldots & z_s^{\ell_s} \\ z_1^{\ell_1} & z_2^{\ell_2 \pm 1} \ldots & z_s^{\ell_s} \\ \vdots & & \\ z_1^{\ell_1} & z_2^{\ell_2} \ldots & z_s^{\ell_s \pm 1} \end{array}$$

If each step is given weight unity, then the aggregate of locations after a single step has the weight:

$$(z_1 + \frac{1}{z_1} + z_2 + \frac{1}{z_2} + \ldots + z_s + \frac{1}{z_s}) z_1^{\ell_1} \ldots z_s^{\ell_s} .$$

Hence, if we start our walk at the origin, then the generating function for the positions after n steps is

$$g_n(z_1,\ldots,z_s) = (z_1 + \frac{1}{z_1} + \ldots + z_s + \frac{1}{z_s})^n .$$

We can now answer some obvious questions: If all directions are equally probable, what is the probability $P_n(0)$ of a random walk returning to the origin after n steps? Since the probability of a step in any given direction is 1/2s, then clearly

$$P_n(0) = \text{coefficient of } z_1^0 \ldots z_s^0 \text{ in } \frac{1}{(2s)^n} g_n(z_1,\ldots,z_s)$$

$$= \text{coefficient of } z_1^0 \cdots z_s^0 \text{ in } \frac{1}{(2s)^n}\left(z_1 + \frac{1}{z_1} + z_2 + \frac{1}{z_2} + \frac{1}{z_2} + \cdots + z_s + \frac{1}{z_s}\right)^n$$

$$= \frac{1}{(2\pi i)^s} \oint \cdots \oint \frac{dz_1}{z_1}\frac{dz_2}{z_2} \cdots \frac{dz_s}{z_s} \frac{\left(z_1 + \frac{1}{z_1} + \cdots + z_s + \frac{1}{z_s}\right)^n}{(2s)^n},$$

where each integral is around a unit circle. If we set $z_j = e^{i\theta_j}$ this becomes

$$P_n(0) = \frac{1}{(2\pi)^s} \int_0^{2\pi} \cdots \int_0^{2\pi} \left(\frac{\cos\theta_1 + \cos\theta_2 + \cdots + \cos\theta_s}{s}\right)^n d\theta_1 \ldots d\theta_s.$$

Notice that since a walk cannot return to the origin after an odd number of steps, $P_n(0) = 0$ unless n is even.

<u>Examples</u>: (i) $s = 1$

$$P_{2n}(0) = \frac{1}{2\pi}\int_0^{2\pi} \cos^{2n}\theta \, d\theta = \frac{1}{2\pi}\frac{1}{2^{2n}}\int_0^{2\pi} (e^{i\theta} + e^{-i\theta})^{2n} d\theta$$

$$P_{2n}(0) = \frac{1}{2^{2n}}\binom{2n}{n}$$

(ii) $s = 2$

$$P_{2n}(0) = \frac{1}{4\pi^2}\int_0^{2\pi}\int_0^{2\pi} \left(\frac{\cos\theta_1 + \cos\theta_2}{2}\right)^{2n} d\theta_1 \, d\theta_2$$

$$= \frac{1}{4\pi^2}\int_0^{2\pi}\int_0^{2\pi} [\cos\tfrac{1}{2}(\theta_1+\theta_2)\cos\tfrac{1}{2}(\theta_1-\theta_2)]^{2n} d\theta_1 \, d\theta_2$$

$$= \frac{1}{4\pi^2}\int_{-\pi}^{\pi}\int_0^{\pi} [\cos\alpha\cos\beta]^{2n} 2 \, d\alpha \, d\beta$$

$$= \left[\frac{1}{2\pi}\int_{-\pi}^{\pi} (\cos\theta)^{2n} d\theta\right]^2 = \frac{1}{4^{2n}}\binom{2n}{n}^2.$$

For $s > 2$ simple exact results are not available. However, the generating function

$$P(x|0) = \sum_{n=0}^{\infty} P_n(0) \frac{x^n}{n!}$$

can be obtained in closed form. We have

$$P(x|0) = \frac{1}{(2\pi)^s} \int_0^{2\pi} \cdots \int_0^{2\pi} \exp[\tfrac{x}{s}(\cos\theta_1 + \cdots + \cos\theta_s)] d\theta_1 \cdots d\theta_s.$$

But

$$\frac{1}{2\pi} \int_0^{2\pi} e^{\alpha\cos\theta} d\theta = I_0(\alpha),$$

where $I_0(\alpha)$ is the zero-th order modified Bessel function

$$I_0(\alpha) = \sum_{k=0}^{\infty} (\tfrac{\alpha}{2})^{2k} \frac{1}{(k!)^2} .$$

We conclude that

$$P(x|0) = (I_0(\tfrac{x}{s}))^s.$$

To use this for large s we may employ the recursion relation

$$I_0(ax)I_0(bx) = \frac{1}{2\pi} \int_0^{2\pi} I_0([2\cos\theta(a^2e^{i\theta} + b^2e^{-i\theta})]^{1/2}x) d\theta.$$

For example

$$(I_0(\tfrac{x}{2}))^2 = \frac{1}{2\pi} \int_0^{2\pi} I_0(x\cos\theta) d\theta$$

$$= \frac{1}{2\pi} \int_0^{2\pi} \sum_{k=0}^{\infty} (\tfrac{x}{2})^{2k} \frac{1}{(k!)^2} (\cos\theta)^{2k} d\theta,$$

but

$$\frac{1}{2\pi} \int_0^{2\pi} (\cos\theta)^{2k} d\theta = \frac{1}{2\pi} \int_0^{2\pi} \sum_{j=0}^{2k} \binom{2k}{j} (\tfrac{1}{2} e^{i\theta})^{2k-j} (\tfrac{1}{2} e^{-i\theta})^j d\theta$$

$$= \binom{2k}{k} \frac{1}{2^{2k}},$$

and it follows at once that $P_{2k}(0) = [\frac{1}{2^{2k}} \binom{2k}{k}]^2$ as before. Continuing in this way, for $s = 3$,

$$P(x|0) = (I_0(\tfrac{x}{3}))^3 = I_0(\tfrac{x}{3})(I_0(\tfrac{x}{3}))^2$$

$$= \frac{1}{2\pi}\int_0^{2\pi} I_0(\tfrac{x}{3}) I_0(\tfrac{2x}{3}\cos\theta)d\theta$$

$$= \frac{1}{4\pi^2}\int_0^{2\pi}\int_0^{2\pi} I_0([2\cos\alpha(\tfrac{1}{9}e^{i\alpha} + \tfrac{4}{9}\cos^2\theta\, e^{-i\alpha})]^{1/2}x)d\theta d\alpha$$

$$= \frac{1}{4\pi^2}\int_0^{2\pi}\int_0^{2\pi}\sum_{k=0}^{\infty}(\tfrac{x}{2})^{2k}\frac{1}{(k!)^2}[\tfrac{2}{9}\cos\alpha(e^{i\alpha}+4\cos^2\theta\, e^{-i\alpha})]^k\, d\theta d\alpha.$$

Hence

$$P_{2k}(0) = \frac{1}{9^k}\frac{1}{2^{2k}}\binom{2k}{k}\frac{1}{4\pi^2}\int_0^{2\pi}\int_0^{2\pi}[2\cos\alpha(e^{i\alpha}+4\cos^2\theta e^{-i\alpha})]^k d\theta d\alpha$$

$$= \frac{1}{6^{2k}}\binom{2k}{k}\frac{1}{4\pi^2}\int_0^{2\pi}\int_0^{2\pi}(1+e^{2i\alpha})^k\sum_{j=0}^{k}\binom{k}{j}4^j\cos^{2j}\theta\, e^{-2i\alpha j}d\theta d\alpha$$

$$= \frac{1}{6^{2k}}\binom{2k}{k}\sum_{j=0}^{k}\binom{k}{j}^2\binom{2j}{j}.$$

Suppose next that $Q_n(0)$ is the probability of returning to the origin for the first time at the n-th step. Clearly,

$$P_{2n}(0) = Q_{2n}(0) + Q_{2n-2}(0)P_2(0) + \dots + Q_2(0)P_{2n-2}(0).$$

Define $P_0(0) = 1$; then $P_{2n}(0) = \sum_{t=0}^{n-1} Q_{2n-2t}(0)P_{2t}(0)$. Hence if we introduce the new generating function

$$P(x) = \sum_{m=0}^{\infty} P_m(0)x^m, \quad Q(x) = \sum_{m=2}^{\infty} Q_m(0)x^m,$$

then we have the relation

$$P(x) - 1 = Q(x)P(x)$$

so that

$$Q(x) = 1 - \frac{1}{P(x)}.$$

Let Q be the probability of ever returning to the origin; then $Q = \sum_{m=1}^{\infty} Q_{2m}(0) =$

$Q(1)$, or

$$Q = 1 - \frac{1}{P(1)} .$$

Therefore

$$P(1) = \infty \quad \text{implies} \quad Q = 1$$

$$P(1) < \infty \quad \text{implies} \quad Q < 1.$$

The probability of a particle ever returning to the origin depends upon the number of ways it can escape from the origin and hence would be expected to decrease as the dimensionality increases. To prove this rigorously, we observe that

$$P(x) = \sum_{m=0}^{\infty} x^m P_m(0) = \sum_{m=0}^{\infty} \frac{x^m}{(2\pi)^s} \int_{-\pi}^{\pi} \cdots \int_{-\pi}^{\pi} \left(\frac{\sum_{i=1}^{s} \cos \theta_i}{s}\right)^m d\theta_1 \ldots d\theta_s$$

$$= \frac{1}{(2\pi)^s} \int_{-\pi}^{\pi} \cdots \int_{-\pi}^{\pi} \frac{d\theta_1 \ldots d\theta_s}{1 - \frac{x}{s} \sum_{j=1}^{s} \cos \theta_j} .$$

In particular

$$P(1) = \frac{1}{(2\pi)^s} \int_{-\pi}^{\pi} \cdots \int_{-\pi}^{\pi} \frac{d\theta_1 \ldots d\theta_s'}{1 - \frac{1}{s} \sum_{j=1}^{s} \cos \theta_j} .$$

We want at the very least to know whether or not $P(1)$ is infinite. The only singularity of the integrand is at $\theta_1 = \ldots = \theta_s = 0$. In the vicinity of this singularity we may use the expansion $\cos \theta \cong 1 - \theta^2/2$. Hence $P(1)$ is infinite if and only if

$$\overline{P} = \frac{1}{(2\pi)^s} \int_{\text{origin}} \cdots \int \frac{d\theta_1 \ldots d\theta_s}{\frac{1}{2s} \sum_{j=1}^{s} \theta_j^2}$$

is infinite. Converting to hyperspherical coordinates

$$d\theta_1 \ldots d\theta_s = r^{s-1} dr\, d\Omega$$

where

$$r^2 = \sum_{j=1}^{s} \theta_j^2$$

we get

$$\overline{P} = \frac{2s\,\Omega}{(2\pi)^s} \int_0^{\varepsilon} r^{s-1} \frac{dr}{r^2} .$$

Hence

$$\overline{P} < \infty \quad \text{for} \quad s \geq 3,$$

$$\overline{P} = \infty \quad \text{for} \quad s = 1,2.$$

We conclude that (Polya):

$$Q = 1 \quad \text{for} \quad s = 1,2,$$

$$Q < 1 \quad \text{for} \quad s \geq 3.$$

It is not difficult to find the asymptotic value of Q for large s. It is most convenient to use the exponential generating function $P(x|0) = \sum_{n=0}^{\infty} P_n(0) \frac{x^n}{n!}$. Since $\int_0^{\infty} e^{-x} \frac{x^n}{n!} dx = 1$, we see that

$$P(1) = \sum_{n=0}^{\infty} P_n(0) = \int_0^{\infty} P(x|0)\, e^{-x} dx = \int_0^{\infty} [I_0(\tfrac{x}{s})]^s\, e^{-x} dx.$$

Observe first that since $I_0(y) \simeq (2\pi y)^{-1/2} e^y$ as $y \to \infty$, $P(1)$ diverges for $s = 1$ or 2 as previously. On the other hand, for large s, $P(1)$ converges and the integrand is monotonically decreasing in x with a rate which increases with s. Thus the asymptotic contribution for large s comes from x around zero. We have

$$(I_0(\tfrac{x}{s}))^s = \exp[s \log I_0(\tfrac{x}{s})]$$

$$= \exp s[\tfrac{1}{4}(\tfrac{x}{s})^2 - \tfrac{1}{64}(\tfrac{x}{s})^4 + \ldots]$$

$$= 1 + \frac{1}{s}\frac{x^2}{4} + \frac{1}{2s^2}\frac{x^4}{16} + \frac{1}{6s^3}\frac{x^6}{64} - \frac{1}{s^3}\frac{x^4}{64} + \ldots ,$$

so that

$$P(1) = 1 + \frac{1}{2}\frac{1}{s} + \frac{3}{4}\frac{1}{s^2} + \frac{3}{2}\frac{1}{s^2} + \ldots .$$

Hence

$$\frac{1}{Q} = P(1)\,\frac{1}{P(1)-1} = 2s - 2 - \frac{3}{2s} + \ldots$$

and we conclude that

$$Q = (2s - 2 - \frac{3}{2s} + \ldots)^{-1} \quad \text{for large } s.$$

2§. General Lattices.

We now consider random walks on regular lattices which may be non-cubical, may involve steps to non-nearest neighbors, or steps with unequal probability. Let the lattice points again be designated by $\{\underset{\sim}{\ell}\}$. Then in the general situation of a translation invariant random walk, the probability of step from $\underset{\sim}{\ell}$ to $\underset{\sim}{\ell}'$ is given by $p(\underset{\sim}{\ell}'-\underset{\sim}{\ell})$. If a vertex $\underset{\sim}{\ell}$ is given the weight $e^{i\underset{\sim}{\ell}\cdot\underset{\sim}{\theta}}$ then the locations with their associated probabilities resulting from a step from $\underset{\sim}{\ell}'$ clearly have the weight

$$\sum_{\underset{\sim}{\ell}} p(\underset{\sim}{\ell}-\underset{\sim}{\ell}')e^{i\underset{\sim}{\ell}\cdot\underset{\sim}{\theta}} = e^{i\underset{\sim}{\ell}'\cdot\underset{\sim}{\theta}} \sum_{\underset{\sim}{\ell}} p(\underset{\sim}{\ell}-\underset{\sim}{\ell}')e^{i(\underset{\sim}{\ell}-\underset{\sim}{\ell}')\cdot\underset{\sim}{\theta}}$$

$$= e^{i\underset{\sim}{\ell}'\cdot\underset{\sim}{\theta}} \sum_{\underset{\sim}{\ell}} p(\underset{\sim}{\ell})e^{i\underset{\sim}{\ell}\cdot\underset{\sim}{\theta}} .$$

In other words, a single step is represented by the transition

$$e^{i\underset{\sim}{\ell}'\cdot\underset{\sim}{\theta}} \to e^{i\underset{\sim}{\ell}'\cdot\underset{\sim}{\theta}}\,\lambda(\underset{\sim}{\theta}),$$

where

$$\lambda(\underset{\sim}{\theta}) \equiv \sum_{\underset{\sim}{\ell}} p(\underset{\sim}{\ell}) e^{i\underset{\sim}{\ell}\cdot\underset{\sim}{\theta}}$$

We conclude that if a random walk starts at the origin, the probability of arriving at $\underset{\sim}{\ell}$ after n steps is given by

$$P_n(\underset{\sim}{\ell}) = \text{coef of } e^{i\underset{\sim}{\ell}\cdot\underset{\sim}{\theta}} \text{ in } (\lambda(\underset{\sim}{\theta}))^n$$

$$= \frac{1}{(2\pi)^s} \int_{-\pi}^{\pi} \cdots \int_{-\pi}^{\pi} [\lambda(\underset{\sim}{\theta})]^n e^{-i\underset{\sim}{\ell}\cdot\underset{\sim}{\theta}} d\theta_1 \ldots d\theta_s .$$

Computations are most easily carried out in terms of generating functions. If we define

$$P(x;\underset{\sim}{\ell}) = \sum_{n=0}^{\infty} P_n(\underset{\sim}{\ell}) x^n$$

$$P(x|\underset{\sim}{\ell}) = \sum_{n=0}^{\infty} P_n(\underset{\sim}{\ell}) \frac{x^n}{n!} ,$$

we have at once from the above that

$$P(x;\underset{\sim}{\ell}) = \frac{1}{(2\pi)^s} \int_{-\pi}^{\pi} \cdots \int_{-\pi}^{\pi} \frac{e^{-i\underset{\sim}{\ell}\cdot\underset{\sim}{\theta}}}{1-x\lambda(\theta)} d\theta_1 \ldots d\theta_s$$

$$P(x|\underset{\sim}{\ell}) = \frac{1}{(2\pi)^s} \int_{-\pi}^{\pi} \cdots \int_{-\pi}^{\pi} e^{-i\underset{\sim}{\ell}\cdot\underset{\sim}{\theta}} e^{x\lambda(\underset{\sim}{\theta})} d\theta_1 \ldots d\theta_s .$$

Examples.

(i) Nearest neighbor random walk on a face centered cubic lattice.

A face centered cubic lattice consists of the vertices of a $2 \times 2 \times 2$ regular cubic lattice together with the centers of all faces. Thus the cube lying between (0,0,0) and (2,2,2) will have the additional vertices (1,1,0), (1,0,1), (0,1,1), (1,1,2), (1,2,1), (2,1,1). Consider any vertex say (0,0,0). Its nearest neighbors will be at a distance of $\sqrt{2}$, namely $(\pm 1,\pm 1,0)$, $(\pm 1,0,\pm 1)$, $(0,\pm 1,\pm 1)$. It follows that for the face centered cubic lattice with equally weighted steps

$$\lambda(\underset{\sim}{\theta}) = \frac{1}{3}(\cos\theta_1 \cos\theta_2 + \cos\theta_1 \cos\theta_3 + \cos\theta_2 \cos\theta_3)$$

(ii) Nearest neighbor random walk on body centered cubic lattice.

A body centered cubic lattice consists of the vertices of a $2 \times 2 \times 2$ regular cubic lattice together with the center of each cube. Thus the cube lying between $(0,0,0)$ and $(2,2,2)$ will have the additional vertex $(1,1,1)$. Consider any vertex, say $(0,0,0)$. Its nearest neighbors will be at a distance $\sqrt{3}$, namely $(\pm 1, \pm 1, \pm 1)$. It follows that

$$\lambda(\underset{\sim}{\theta}) = \cos\theta_1 \cos\theta_2 \cos\theta_3.$$

The preceding example is one in which $\lambda(\underset{\sim}{\theta})$ is a product of lower dimensional λ's:

$$\lambda(\theta_1,\ldots,\theta_s) = \lambda'(\theta_1,\ldots,\theta_t)\lambda''(\theta_{t+1},\ldots,\theta_s)$$

Under the conditions $P_n(\underset{\sim}{\ell})$ decomposes the same way

$$P_n(\ell_1,\ldots,\ell_s) = \frac{1}{(2\pi)^s}\int_{-\pi}^{\pi}\cdots\int_{-\pi}^{\pi}[\lambda(\theta_1,\ldots,\theta_s)]^n e^{-i\sum_{j=1}^{s}\ell_j\theta_j}\,d\theta_1\ldots d\theta_s$$

$$= P'_n(\ell_1,\ldots,\ell_t)P''_n(\ell_{t+1},\ldots,\ell_s)$$

where

$$P'_n(\ell_1,\ldots,\ell_t) = \frac{1}{(2\pi)^t}\int_{-\pi}^{\pi}\cdots\int_{-\pi}^{\pi}[\lambda'(\theta_1,\ldots,\theta_t)]^n e^{-i\sum_{j=1}^{t}\ell_j\theta_j}\,d\theta_1\ldots d\theta_t$$

$$P''_n(\ell_{t+1},\ldots,\ell_s) = \frac{1}{(2\pi)^{s-t}}\int_{-\pi}^{\pi}\cdots\int_{-\pi}^{\pi}[\lambda''(\theta_{t+1},\ldots,\theta_s)]^n \cdot e^{-i\sum_{j=t+1}^{s}\ell_j\theta_j}\,d\theta_{t+1}\ldots d\theta_s,$$

e.g. since $\lambda(\theta) = \cos\theta$ for the 1-dimensional random walk the body center cubic probabilities are the products of the component 1-dimensional random walk probabilities:

$$P_{2n}(0) = [\frac{1}{2^{2n}} \binom{2n}{n}]^3.$$

This situation occurs whenever the steps are Cartesian products of lower dimensional steps. Thus the body centered cubic steps are the Cartesian cubes of the 1-dimensional steps (± 1). Note too that the square of the step (± 1) is the set $(\pm 1, \pm 1)$ which generates the square lattice at a 45° angle. This is why the square lattice probabilities are so simple.

B. One Dimensional Lattices.

1§. The Ballot Problem.

Prototype: A wholesale house does business only in \$100 units, one transaction per customer per month. All bills are paid the last day of the month. On this day, n creditors and n debtors appear one at a time. The house has no cash to start with. What is the probability P that some creditor will have to wait to be paid?

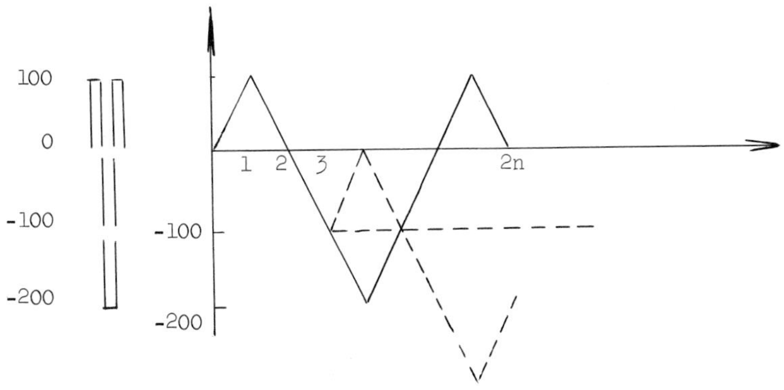

The cash on hand can be regarded as the position of a one dimensional random walk which starts at the origin. It is convenient to represent the process on the time axis as well, as in the diagram above. We shall solve the problem using the reflection principle. Any path reaching -100 corresponds to a creditor having to wait. Now reflect the path about -100 after the first time it reaches -100; there results a path from (0,0) to (2n,-200), and in fact there is clearly a 1-1 correspondence between paths from (0,0) to (2n,0) which reach -100 and all paths from (0,0) to (2n,-200). A path from (0,0) to (2n,-200) must consist of n-1 +100's and n+1 -100's; there are $\binom{2n}{n+1}$ such paths. On the other hand there are $\binom{2n}{n}$ paths consisting of n +100's and n -100's. We conclude that

$$P = \frac{\binom{2n}{n+1}}{\binom{2n}{n}} = \frac{n}{n+1} .$$

In the usual statement of the ballot problem, $n = a+b$ votes are cast in sequence, a for candidate A, b for candidate B. At the r-th step, A has α_r votes, B has β_r. What is the probability of some relation between α_r and β_r always holding, holding exactly j times, etc,? For example, in the above problem $a = b$ and we were interested in the probability of a strict lead for B, $\beta_r - \alpha_r > 0$, for at least one r.

We will now consider the class of problems which inquire only as to the properties of

$$s_r = \alpha_r - \beta_r .$$

Let us define the state at the r-th step by an infinite vector u_r with a unit only at the s_r-th position.

$$u_{r,i} = \delta_{s_r,i}, \text{ where } u_{r,i} \text{ is the i-th component of } u_r.$$

Clearly then

$$u_r = E^{\sigma_r} u_{r-1}$$

where $\sigma_r = \pm 1$ and the matrix elements

$$E_{ij} = \begin{cases} 1 & j = i-1 \\ 0 & j \neq i-1 \end{cases}$$

$$E^{-1}_{ij} = \begin{cases} 1 & j = i+1 \\ 0 & j \neq i+1 \end{cases} .$$

That is E^{-1} raises the location of the unit by 1 and E lowers it by 1. For example, if the process starts at the origin then the sum of all possible states at the r-th step with equal weight is given by

$$\Sigma\{u_{r,i}\} = ((E+E^{-1})^r u_0)_i = \sum_{j=0}^{r} \binom{r}{j} (E^{r-2j} u_0)_i = \binom{r}{\frac{1}{2}(r-i)}.$$

Now suppose that a state with a unit at the j-th position is given a weight W_j, and define the weight of a chain as the product of the weights, excluding the initial position. Hence

$$W(\sigma_1,\ldots,\sigma_n | u_0, u) = (u, WE^{\sigma_n} WE^{\sigma_{n-1}} \ldots WE^{\sigma_1} u_0)$$

is the weight from u_0 to u specified by the steps $\sigma_1,\ldots,\sigma_n$. The matrix $W = (W_{ij})$ is defined here by

$$W_{ij} \equiv W_j \delta_{ij}.$$

It follows that the total weight of all chains of length n between u_0 and u is given by

$$W_n(u_0, u) = (u, W(E+E^{-1})W(E+E^{-1}) \ldots (E+E^{-1}) u_0)$$

$$= (u, [W(E+E^{-1})]^n u_0).$$

Finally, we may define a generating function with respect to n,

$$W(u_0, u, \lambda) = \sum_{n=0}^{\infty} \lambda^n W_n(u_0, u) = (u, [I - \lambda W(E+E^{-1})]^{-1} u_0)$$

the last sum converging whenever $\lambda < |\lambda_{max}\{W(E+E^{-1})\}|^{-1}$.

Example. Find the number of chains $Z_{a,b}(t)$ with exactly t tie positions ($s_r = 0$, $r \neq 0$) in the balloting.

To recognize a tie position we use a multiplicative weight of z for the position $j = 0$, a weight of unity for all other positions, and then look for the coefficient of z^t. Hence

$$Z_{a,b}(t) = \text{coef of } \lambda^{a+b} z^t \text{ in } v_{a-b},$$

where v_{a-b} is (a-b)-th component of v and

$$v - \lambda W(z)(E+E^{-1})v = u_0$$

where $W_j(z) = z^{\delta_{j,0}}$.

Thus we have to solve the difference equation

$$v_j - \lambda(v_{j+1} + v_{j-1}) = 0 \quad \text{for} \quad j \neq 0$$

with boundary condition

$$v_0 - \lambda z(v_1 + v_{-1}) = 1.$$

The difference equation has the solution

$$v_j = v_0 x^{|j|} \quad \text{where} \quad 1 - \lambda(x + \frac{1}{x}) = 0$$

and $x \to 0$ as $\lambda \to 0$. From the boundary condition

$$v_0 - 2\lambda z v_0 x = 1$$

or

$$v_0 = \frac{1}{1 - 2\lambda z x}.$$

Hence

$$v_j = \frac{x^{|j|}}{1 - 2\lambda z x} = \sum_{t=0}^{\infty} 2^t \lambda^t z^t x^{t+|j|}.$$

Our remaining task is to evaluate $x^{t+|j|}$. Since $x = 0+\lambda(1+x^2)$ then according to the Lagrange inversion theorem:

$$x^{t+|j|} = 0^{t+|j|} + \sum_{i=1}^{\infty} \frac{1}{i!} \frac{d^{i-1}}{dx^{i-1}} (\lambda\,(1+x^2))^i (t+|j|)x^{t-1+|j|} \bigg|_{x=0}$$

for $t + |j| > 0$

$$= (t+|j|) \sum_{i=1}^{\infty} \lambda^i \frac{1}{i} \text{ coef } x^{i-1} \text{ in } x^{t-1+|j|}(1+x^2)^i$$

$$= (t+|j|) \sum_{i=1}^{\infty} \lambda^i \frac{1}{i} \text{ coef } x^{i-t-|j|} \text{ in } \sum_{s=0}^{i} \binom{i}{s} x^{2s}$$

$$= (t+|j|) \sum_{s=0}^{\infty} \lambda^{2s+t+|j|} \frac{1}{t+|j|+2s} \binom{t+|j|+2s}{s}.$$

Therefore

$$Z_{a,b}(t) = \text{coef of } \lambda^{a+b} \text{ in}$$

$$2^t \sum_{s=0}^{\infty} \lambda^{2t+|a-b|+2s} \frac{t+|a-b|}{t+|a-b|+2s} \binom{t+|a-b|+2s}{s}$$

or if $a \geq b$

$$Z_{a,b}(t) = 2^t \frac{a-b+t}{a+b-t} \binom{a+b-t}{b-t}.$$

2§. One Dimensional Lattice Gas.

In the above, rather than an algebraic specification of the development of a chain, we have used the transfer matrix $(E+E^{-1})$ to make each step, and a weight $W(z)$ to assess it. The two of course are intimately related. As a second field of application for the transfer matrix, consider a periodically bounded finite discrete gas (the particles are confined to the vertices of a lattice wound into a ring). The occupation is specified by $\nu_\ell = 0$ or 1 at each integer $\ell = 1,\ldots,\Omega$ and the interaction is $E_{\ell,\ell+1} = \phi\, \nu_\ell \nu_{\ell+1}$ between nearest neighbors only. Then according to equilibrium statistical mechanics, if $\beta = 1/\kappa T$, where T is the temperature, the pair $(\ell,\ell+1)$ has the weight

$$W_{\ell,\ell+1} = e^{-\beta\phi\, \nu_\ell \nu_{\ell+1}},$$

and a lattice gas configuration has the weight

$$W(\nu_1,\ldots,\nu_\Omega) = \exp - \sum_{\ell=1}^{\Omega} \beta\phi\nu_{\ell-1}\nu_\ell,$$

where $\nu_0 \equiv \nu_\Omega$.

We may now consider the restricted number problem: There are exactly N particles, or $\sum_{\ell=1}^{\Omega} \nu_\ell = N$. Typical questions to ask are the total weight of all configurations

$$W_N(\Omega) = \sum_{\nu_\ell=0,1 \,|\, \sum_{\ell=1}^{\Omega} \nu_\ell = N} W(\nu_1,\ldots,\nu_\Omega),$$

the corresponding free energy defined as:

$$F_N(\Omega) = -\frac{1}{\beta} \log W_N(\Omega).$$

($F_N(\Omega) = \phi$ when there is only one occupied pair), and the two site "correlation" function defined as:

$$g_N(\ell|\Omega) = \frac{\langle \nu_1 \nu_{\ell+1} \rangle}{\langle \nu_1 \rangle \langle \nu_{\ell+1} \rangle}.$$

A more special problem of importance is obtained by taking the limit $\beta\phi \to +\infty$. Hence $W_{\ell,\ell+1} = 1 - \nu_\ell \nu_{\ell+1}$ is non-vanishing only if one of the sites ℓ, $\ell+1$ is unoccupied, and $W_N(\Omega)$ = number of configurations with no two adjacent sites occupied. In this case $N \leq \frac{1}{2}\Omega$, the "close-packed" configuration. We can also consider the dual problem of occupying line segments with the restriction that two adjacent line segments not be simultaneously occupied. An occupied line segment with its two vertices is called a dimer, and so we are then asking for the total number of ways of inserting dimers on a lattice. If $N = \frac{1}{2}\Omega$, this is the number of ways of completely filling the lattice with dimers, which is trivial in 1-dimensional (the answer is 2) but quite difficult for higher dimensional lattices.

The lattice gas model is also identical with a weighted 1-dimensional random walk. Define $\sigma_\ell = 2\nu_\ell - 1 = \pm 1$. We then have N (+1)'s and Ω-N (-1)'s, which may be interpreted as successive steps on the axis. This is a restricted random walk, since it must terminate at $2N-\Omega$ at the Ω step. The weight now is

$$\prod_{\ell=1}^{\Omega} \exp - \frac{1}{4}\beta\phi(1+\sigma_\ell+\sigma_{\ell-1}+\sigma_\ell\sigma_{\ell-1}) = e^{-\beta\phi(N-\frac{\Omega}{4})} \prod_{\ell=1}^{\Omega} e^{-\frac{1}{4}\beta\phi\sigma_\ell\sigma_{\ell-1}}.$$

For the full weight, it is convenient to drop the first factor, which is a constant, so that we define

$$W_N(\Omega) = \sum^{\Omega}_{\sigma_\ell = \pm 1 |\ \sum_{\ell=1}^{\Omega} \sigma_\ell = 2N-\Omega} \exp - \frac{\beta\phi}{4} \sum_{\ell=1}^{\Omega} \sigma_\ell\sigma_{\ell-1}.$$

We may also consider the corresponding unrestricted random walk, with

$$W(\Omega) = \sum_{\sigma_\ell = \pm 1} \exp - \frac{1}{4}\beta\phi \sum_{\ell=1}^{\Omega} \sigma_\ell\sigma_{\ell-1}.$$

Here the expected length of the walk $\langle \sum_{\ell=1}^{\Omega} \sigma_\ell \rangle = 0$, since the weight $W(\sigma_1,\dots,\sigma_\Omega)$ is an even function of the vector $(\sigma_1,\dots,\sigma_\Omega)$ whereas $\sum_{\ell=1}^{\Omega} \sigma_\ell$ is an odd function. Note too that $\lim_{\Omega\to\infty} \frac{1}{\Omega} \langle (\sum_{\ell=1}^{\Omega} \sigma_\ell)^2 \rangle \to 0$.

The unrestricted random walk is also identical with the (field-free) one-dimensional Ising Model: $\frac{1}{2}\sigma$ then refers to the "spin" $\pm 1/2$, and only neighboring spins interact.

Let us return now to the lattice gas. For the unrestricted gas, we choose the same weight (to within the constant $e^{-\frac{1}{4}\phi\beta\Omega}$) as the unrestricted random walk:

$$W(\Omega) = \sum_{\nu_\ell=0,1} e^{-\beta\phi \sum_{\ell=1}^{\Omega} \nu_\ell\nu_{\ell-1}} e^{\beta\phi \sum_{\ell=1}^{\Omega} \nu_\ell}.$$

It is clear then that $\langle 2N-\Omega\rangle = 0$ or $\langle N\rangle = \frac{1}{2}\Omega$: the mean particle density is automatically 1/2. We solve the problem of total weight and pair correlation function by setting up the transfer matrix for each step of the corresponding random walk. Thus if

$$\overline{W}_{\nu\nu'} = e^{-\beta\phi\nu\nu'}, \quad Z_{\nu\nu'} = e^{\beta\phi\nu}\delta_{\nu,\nu'}$$

then

$$(\overline{W}Z)_{\nu\nu'} = e^{-\beta\phi\nu\nu'}\, e^{\beta\phi\nu'},$$

and we have

$$W(\Omega) = \mathrm{Tr}\,(\overline{W}Z)^{\Omega}$$, where Tr is defined as the trace.

Similarly, if we define $V_{\nu\nu'} = \nu\delta_{\nu,\nu'}$ $\nu,\nu' \in \{0,1\}$, it easily follows that

$$\langle \nu_1 \nu_{\ell+1} \rangle = \mathrm{Tr}(V(\overline{W}Z)^{\ell} V(\overline{W}Z)^{\Omega-\ell})/\mathrm{Tr}(\overline{W}Z)^{\Omega}.$$

Somewhat more simply, if we introduce the symmetric matrix

$$W = Z^{1/2}\,\overline{W}\,Z^{1/2} = \begin{pmatrix} 1 & e^{-\frac{1}{2}\beta\phi} \\ e^{-\frac{1}{2}\beta\phi} & 1 \end{pmatrix}$$

then

$$W(\Omega) = \mathrm{Tr}\,W^{\Omega}, \quad \langle \nu_1 \nu_{\ell+1} \rangle = \mathrm{Tr}(VW^{\ell}VW^{\Omega-\ell})/\mathrm{Tr}\,W^{\Omega}.$$

Since W has the eigenvalues:

$$\lambda_1 = 1 + e^{-\frac{1}{2}\beta\phi}, \quad \lambda_2 = 1 - e^{-\frac{1}{2}\beta\phi}$$

and eigenvectors:

$$u_1 = 2^{-1/2}\binom{1}{1}, \quad u_2 = 2^{-1/2}\binom{1}{-1},$$

we can evaluate W^n at once as

$$W^n = \lambda_1^n\, u_1 u_1' + \lambda_2^n\, u_2 u_2',$$ where u' is the transpose of u.

Thus

$$\mathrm{Tr}(W^{\Omega}) = \lambda_1^{\Omega} + \lambda_2^{\Omega}$$

and

$$\mathrm{Tr}(VW^{\ell}VW^{\Omega-\ell}) = u_1'V(\lambda_1^{\ell}u_1u_1' + \lambda_2^{\ell}u_2u_2')V\lambda_1^{\Omega-\ell}u_1$$

$$+ u_2'V(\lambda_1^{\ell}u_1u_1' + \lambda_2^{\ell}u_2u_2')V\lambda_2^{\Omega-\ell}u_2$$

$$= \frac{1}{4}\lambda_1^{\Omega} + \frac{1}{4}\lambda_1^{\Omega-\ell}\lambda_2^{\ell} - \frac{1}{4}\lambda_1^{\ell}\lambda_2^{\Omega-\ell} - \frac{1}{4}\lambda_2^{\Omega}$$

$$= \frac{1}{4}(\lambda_1^{\Omega-\ell} - \lambda_2^{\Omega-\ell})(\lambda_1^{\ell} + \lambda_2^{\ell}).$$

In other words,

$$F(\Omega) = -\frac{1}{\beta}[\Omega \log \lambda_1 + \log(1 + (\frac{\lambda_2}{\lambda_1})^{\Omega})]$$

$$g(\ell|\Omega) = [1 + (\frac{\lambda_2}{\lambda_1})^{\ell}] \frac{1 - (\frac{\lambda_2}{\lambda_1})^{\Omega-\ell}}{1 + (\frac{\lambda_2}{\lambda_1})^{\Omega}},$$

and in the limit, $\Omega \to \infty$, these become

$$F(\Omega) \to -\frac{\Omega}{\beta}\log \lambda_1,$$

$$g(\ell|\Omega) \to 1 + (\frac{\lambda_2}{\lambda_1})^{\ell}.$$

C. Two Dimensional Lattices.

1§. Counting Figures on a Lattice, General Algebraic Approach.

We shall be concerned with the dimer problem and its extensions - the number of ways, $\mathcal{N}$, a set of vertex-occupying figures can completely cover a given lattice L without overlapping. Later, we will examine the corresponding weighted configurations, as in statistical mechanics. To be very general, suppose we consider a collection $f = \{\phi\}$ of figures ϕ on a lattice of vertices $\{\alpha\}$. Each ϕ can be given a weight $\prod_{\alpha\in\phi} x_\alpha$, so that the weight for either having or not having the specific figure ϕ is $(1 + \prod_{\alpha\in\phi} x_\alpha)$. Hence

$$\mathcal{N}\{\phi,\alpha\} = \text{coef of } \prod_{\alpha\in L} x_\alpha \text{ in } \prod_{\phi\in f}(1 + \prod_{\alpha\in\phi} x_\alpha)$$

$$= \text{coef of } \prod_{\alpha\in L} x_\alpha \text{ in } \exp \sum_{\phi\in f} \prod_{\alpha\in\phi} x_\alpha .$$

For example, for dimers (2 vertex links or dominos) on a 2M by 2N square lattice,

$$\mathcal{N} = \text{coef} \prod_{i,j} x_{ij} \text{ in } \exp \sum_{i=1}^{M} \sum_{j=1}^{N} [x_{2i-1,2j-1}\, x_{2i,2j-1}$$

$$+ x_{2i-1,2j-1}\, x_{2i-1,2j} + x_{2i,2j}\, x_{2i-1,2j} + x_{2i,2j}\, x_{2i,2j-1}]$$

This can also be written in terms of a determinant by the inverse Master theorem approach, but is not easy to deal with. Also, since the dimer problem can be written in terms of allowed permutations - the terminus on the B sublattice of a dimer originating on the A sublattice - it can be written directly as a determinantal evaluation, which is even useful for small lattices. The progress which has been made, however, has been by avoidance of two dimensional topological considerations. For this purpose one chooses composite basic figures, e.g. a basic figure might be a row on a square lattice and its associated dimers. The composite basic figures can then be linearly ordered, as on the one dimensional lattice, and the transfer matrix approach is again applicable.

An alternative generating function technique is figure-oriented rather than vertex-oriented. Suppose that X_ϕ is the weight of a figure of p_ϕ vertices. Then $\sum_{\phi\ni\alpha} X_\phi$ counts the possible figures at a given vertex α. If we construct $\prod_\alpha (\sum_{\phi\ni\alpha} X_\phi)$, then each figure must be present either p_ϕ times or not at all, since each figure which occurs in a non-overlapping covering will occur at p_ϕ vertices. It follows that

$$\mathcal{N}\{\phi,\alpha\} = \text{coef of} \prod_{\phi\in f} X_\phi^{p_\phi} \text{ in} \prod_{\alpha\in L} (\sum_{\phi\ni\alpha} X_\phi) \prod_{\phi\in f} (1+X_\phi^{p_\phi}).$$

This expression can be evaluated in numerous ways e.g. by the inverse Master theorem technique, for $p_\phi = 2$,

$$\mathcal{N}\{\phi,\alpha\} = \int_{-\infty}^{\infty} \cdots \int_{-\infty}^{\infty} \exp(-\tfrac{1}{2}\sum_{\phi\in f} X_\phi^2) \prod_{\alpha\in L}(\sum_{\phi\ni\alpha} X_\phi) \prod_{\phi\in f} \frac{1}{(2\pi)^{1/2}}\, dX_\phi,$$

which can be applied to the dimer problem. Other more powerful techniques will be described in due course.

2§. The Dimer Problem - Transfer Matrix Method. (E. Lieb)

In how many ways can we fill a $2M$ by $2N$ square lattice with non-overlapping dimers? A dimer is a horizontal or vertical bond which occupies two vertices. In this section we shall consider the case of periodic boundary conditions which means that the first column is to be identified with the $2M+1$ th column and the first row with the $2N+1$ th row.

As mentioned above we may specify a dimer arrangement by listing the type of dimer at each vertex i.e., up, down, to the left or the right. This introduces the possibility of two dimensional topological problems. To avoid these, we artificially construct composite vertices which can be ordered in a one dimensional space. These composite vertices are simply the rows of the lattice. The configurations of a given row u form the base vectors of the space in which our computations will be performed. As we shall soon see, it is only necessary to determine whether or not there is a vertical dimer from any given vertex of a row. From this point of view there are then 2^{2M} base vectors or configurations of a given row. To make this explicit, a vertex j with an upward vertical dimer will be specified by the unit vector $\binom{1}{0}$, or if not occupied by an upward vertical dimer, by $\binom{0}{1}$. The configuration of u is then given by the Cartesian product (juxta-position rather than multiplication e.g.

$$(a_1,\ldots,a_s) \times (b_1,\ldots,b_s) = (a_1b_1,\ldots,a_1b_s,a_2b_1,\ldots,a_2b_s,\ldots))$$

$$u = u(1) \otimes u(2) \otimes \cdots \otimes u(2M)$$

where

$$u(j) = \begin{cases} \binom{1}{0} & \text{if } j \text{ is occupied by a dimer to its upper neighbor} \\ \binom{0}{1} & \text{if } j \text{ is not occupied by a dimer to its upper neighbor} \end{cases}$$

We note that the configurations of u form an orthonormal basis for the 2^{2M} dimensional vector space which they generate.

Suppose now that the bottom row, which we shall call the first, has the configuration u_1. We then define the transfer matrix $T(w,u)$ between a row u and its upper neighbor w by the condition

$$T(w,u) = \begin{cases} 1 & \text{if } w \text{ and } u \text{ are consistent configurations} \\ 0 & \text{otherwise} \end{cases}$$

(by consistent we mean that the designated pair of rows can be continued to a legitimate configuration of the lattice). The quantity

$$Tu_1 = \sum_{u_2} T(u_2,u_1)u_2$$

is then clearly the sum of all possible configurations of the second row given that the first row is u_1. Similarly

$$T^2u_1 = T(\sum_{u_2} T(u_2,u_1)u_2) = \sum_{u_3} \sum_{u_2} T(u_2,u_1)T(u_3,u_2)u_3$$

is the sum of all possible configurations of the third row given that the first is u_1. Iterating this process we see that the sum of all possible configurations of the $2N+1$ th row is given by $T^{2N}u_1$. But according to periodic boundary conditions the configuration of the $2N+1$ th row must be u_1 itself. The number of possible configurations corresponding to u_1 is therefore the coefficient of u_1 in $T^{2N}u_1$ or by orthonormality

$$\mathcal{N}(u_1) = (u_1, T^{2N}u_1).$$

Summing over all u_1 we conclude that

$$\mathcal{N} = \mathrm{Tr}(T^{2N}).$$

Let us evaluate the transfer matrix T. Suppose that we are given a row configuration u and ask for those upper neighbor configurations w which are consistent with u. Clearly, u will interfere with w only by virtue of its vertical dimers which terminate at w. This is why the configuration as we have defined it is sufficient to determine consistency. The row w is characterized first of all by its horizontal dimers. Suppose $S = \{\alpha\}$ is the set of adjacent vertex pairs on which the horizontal dimers of w are placed. A dimer can be placed on w at $(i,i+1)$

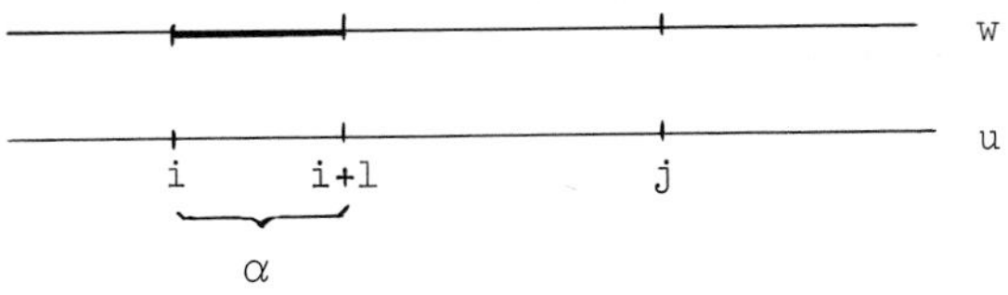

only if there are no vertical dimers from u at $(i,i+1)$ i.e. only if the partial configuration of u for this pair is $\binom{0}{1}\otimes\binom{0}{1}$. Thus with respect to this pair the transfer matrix T must yield 0 for each configuration except $\binom{0}{1}\otimes\binom{0}{1}$, and since the $(i,i+1)$ pair on w cannot then be occupied by vertical dimers, T must take $\binom{0}{1}\otimes\binom{0}{1}$ into itself. The projection which has these characteristics is clearly given by

$$\overline{h}_\alpha = I\otimes I\otimes\cdots\otimes I\otimes\begin{pmatrix}0&0\\0&1\end{pmatrix}\otimes\begin{pmatrix}0&0\\0&1\end{pmatrix}\otimes I\otimes\cdots\otimes I,$$

$$1 \quad 2 \quad \cdots \qquad i \qquad\quad i+1 \qquad\quad \cdots \quad 2M$$

since $\begin{pmatrix}0&0\\0&1\end{pmatrix}\binom{0}{1} = 1\binom{0}{1}$, $\begin{pmatrix}0&0\\0&1\end{pmatrix}\binom{1}{0} = 0\binom{1}{0}$. Consider next a vertex j on w which is not occupied by a horizontal dimer. Its configuration is completely determined by that of the corresponding vertex j on u: a vertical dimer on u means that there cannot be one on w, an absent vertical dimer on u requires a vertical dimer at j on w (since it is not occupied by horizontal dimer). The transfer matrix must then reverse the configuration of j, and the operation which carries this out is

$$v_j = \underset{1}{I} \otimes \underset{2}{I} \otimes \underset{\cdots}{\cdots} \otimes I \otimes \underset{j}{\begin{pmatrix} 0 & 1 \\ 1 & 0 \end{pmatrix}} \otimes I \otimes \underset{\cdots}{\cdots} \otimes \underset{2M}{I}.$$

We conclude at once that

$$T = \sum_s \prod_{\alpha \in s} \bar{h}_\alpha \prod_{j \in \bar{s}} v_j.$$

This expression may be simplified in several ways. First we note that

$$\bar{h}_{i,i+1} = h_{i,i+1} v_i v_{i+1}$$

where

$$h_{i,i+1} = I \otimes I \otimes \cdots \otimes I \otimes \begin{pmatrix} 0 & 0 \\ 1 & 0 \end{pmatrix} \otimes \begin{pmatrix} 0 & 0 \\ 1 & 0 \end{pmatrix} \otimes I \otimes \cdots \otimes I.$$

Hence

$$T = \sum_s \prod_{\alpha \in s} h_\alpha \prod_{j=1}^{2M} v_j.$$

According to our method of construction, the product over α includes only non-overlapping adjacent-element pairs. However, from the fact that $\begin{pmatrix} 0 & 0 \\ 1 & 0 \end{pmatrix}\begin{pmatrix} 0 & 0 \\ 1 & 0 \end{pmatrix} = \begin{pmatrix} 0 & 0 \\ 0 & 0 \end{pmatrix}$, we see that

$$h_\alpha h_\beta = 0 \quad \text{if} \quad \alpha \quad \text{and} \quad \beta \quad \text{overlap.}$$

Thus the product over α may be extended to include any set of adjacent pairs whose union is s, and we have

$$T = \prod_{\alpha=(1,2)}^{(2M,1)} (I + h_\alpha) \prod_{j=1}^{2M} v_j$$

$$= (\exp \sum_{i=1}^{2M} h_{i,i+1}) \prod_{j=1}^{2M} v_j.$$

Finally it will be convenient to consider not T but T^2. Now $v_j v_j = I_{2M \times 2M}$ and

$$v_j v_{j+1} h_{j,j+1} v_j v_{j+1} = I \otimes I \otimes \cdots \otimes I \otimes \begin{pmatrix} 0 & 1 \\ 0 & 0 \end{pmatrix} \otimes \begin{pmatrix} 0 & 1 \\ 0 & 0 \end{pmatrix} \otimes I \otimes \cdots \otimes I.$$

We may write

$$h_{j,j+1} = \sigma_j^- \sigma_{j+1}^-$$

$$v_j v_{j+1} h_{j,j+1} v_j v_{j+1} = \sigma_j^+ \sigma_{j+1}^+$$

where

$$\sigma_j^- = I \otimes \cdots \otimes I \otimes \begin{pmatrix} 0 & 0 \\ 1 & 0 \end{pmatrix} \otimes I \otimes \cdots \otimes I$$

$$\sigma_j^+ = I \otimes \cdots \otimes I \otimes \begin{pmatrix} 0 & 1 \\ 0 & 0 \end{pmatrix} \otimes I \otimes \cdots \otimes I,$$

and in this notation we have

$$T^2 = \exp \sum_{j=1}^{2M} (\sigma_j^- \sigma_{j+1}^-) \exp \sum_{j=1}^{2M} (\sigma_j^+ \sigma_{j+1}^+),$$

a manifestly self adjoint operator.

In the asymptotic limit $N \to \infty$,

$$\mathcal{N} = \mathrm{Tr}(T^2)^N \simeq (\lambda_{max}(T^2))^N,$$

so that we only have to find the maximum eigenvalue of T^2. Let us first use an approximation common in spin-wave theory. Since $\sum_{j=1}^{2M} \sigma_j^- \sigma_{j+1}^-$ with periodic boundary conditions is translation invariant, it can be diagonalized by a Fourier transformation. On the finite space of the row vertices this transformation is

$$\sigma_j^- = \frac{1}{(2M)^{1/2}} \sum_{k=k_0}^{2Mk_0} e^{ijk} S_k^- \quad \text{where} \quad k_0 = \frac{2\pi}{2M}.$$

Then indeed

$$\begin{aligned} \sum_j \sigma_j^- \sigma_{j+1}^- &= \frac{1}{2M} \sum_{jk\ell} e^{ijk} e^{i(j+1)\ell} S_k^- S_\ell^- \\ &= \sum_{k=k_0}^{2Mk_0} e^{-ik} S_k^- S_{-k}^- \quad \text{since} \quad \sum_j e^{ij(k+\ell)} = 2M\delta_{k+\ell,0}. \end{aligned}$$

In the same way

$$\sum_j \sigma_j^+ \sigma_{j+1}^+ = \sum_{k=k_0}^{2Mk_0} e^{-ik} S_k^+ S_{-k}^+ .$$

It is clear that the commutator

$$[S_k^-, S_\ell^-] \equiv S_k^- S_\ell^- - S_\ell^- S_k^- = 0,$$

as well as $[S_k^+, S_\ell^+] = 0$.

On the other hand

$$[S_k^+, S_\ell^-] = \frac{1}{2M} \sum_{j,j'} e^{ijk} e^{ij'\ell} [\sigma_j^+, \sigma_{j'}^-]$$

$$= \frac{1}{2M} \sum_j e^{ij(k+\ell)} [\sigma_j^+, \sigma_j^-]$$

$$= \frac{1}{2M} \sum_j e^{ij(k+\ell)} I \otimes I \otimes \ldots \otimes I \otimes \begin{pmatrix} 1 & 0 \\ 0 & -1 \end{pmatrix} \otimes I \otimes \ldots \otimes I$$

$$\equiv \frac{1}{2M} \sum_{j=1}^{2M} a_j e^{i\phi_j} .$$

If as $M \to \infty$ the amplitudes a_j are bounded and the phases ϕ_j are random then $[S_k^+, S_\ell^-] \to 0$ as well. Thus all $S_k^{\pm}$ commute and we have

$$T^2 = \exp \sum_{k=k_0}^{Mk_0} 2(S_k^- S_{-k}^- + S_k^+ S_{-k}^+) \cos k.$$

Since all of these quantities commute, the maximum eigenvalue of T^2 can be found if we know the range of variation of

$$Q_k = S_k^- S_{-k}^- + S_k^+ S_{-k}^+ .$$

Now

$$S_k^+ S_{-k}^- + S_k^- S_{-k}^+ = \frac{1}{2M} \sum_{j,j'} e^{-i(j-j')k} (\sigma_j^- \sigma_{j'}^+ + \sigma_j^+ \sigma_{j'}^-)$$

$$= I_{2M \times 2M} + \frac{2}{2M} \cos k \sum_j (\sigma_j^- \sigma_{j+1}^+) + \frac{2}{2M} \cos 2k \sum_j \sigma_j^- \sigma_{j+2}^+ + \ldots$$

(since $\sigma_j^-\sigma_j^+ + \sigma_j^+\sigma_j^- = I_{2M \times 2M}$), so that if the previous random phase argument applies we have

$$I_{2M \times 2M} = S_k^+ S_{-k}^- + S_k^- S_{-k}^+.$$

To find the numerical range of Q_k we observe that

$$Q_k + I_{2M \times 2M} = (S_k^+ + S_k^-)(S_{-k}^+ + S_{-k}^-) \geq 0$$

$$-Q_k + I_{2M \times 2M} = (S_k^+ - S_k^-)(S_{-k}^+ - S_{-k}^-) \geq 0$$

whence

$$-1 \leq Q_k \leq +1.$$

We conclude that

$$\lambda_{max}(T^2) = \exp \sum_{k=k_0}^{Mk_0} 2|\cos k|,$$

or carrying out the limits $M \to \infty$, $N \to \infty$,

$$\mathscr{N} \simeq (\lambda_{max}(T^2))^N = \exp \frac{N}{k_0} \int_{k_0}^{Mk_0} 2|\cos k|\, dk$$

$$= \exp \frac{2N}{k_0} \int_0^{\pi/2} 2 \cos k \, dk$$

$$= \exp \frac{1}{\pi} (2M)(2N).$$

The above result is an approximation because the variables $S_k^{\pm}$ do not really commute even when $M \to \infty$ and their domains of variation are inextricably intertwined. Expressing this somewhat differently, the transformation from the $\sigma_k^{\pm}$ to the $S_k^{\pm}$ is not unitary and thus does not retain the simplicity of the structure of the $\sigma_k^{\pm}$. Both problems may be avoided by a preliminary transformation from the $\sigma_k^{\pm}$ to an essentially equivalent set of anticommuting operators (The Paulion to Fermion Transformation). This is done by appending a sign which depends multiplicatively on the states of the preceding vertices. We define

$$a_j \equiv \prod_{i=1}^{j-1} (-\sigma_i^z)\sigma_j^-, \quad a_j^* \equiv \prod_{i=1}^{j-1} (-\sigma_i^z)\sigma_j^+$$

where

$$\sigma_i^z = I \otimes I \otimes \ldots \otimes I \otimes \begin{pmatrix} 1 & 0 \\ 0 & -1 \end{pmatrix} \otimes I \otimes \ldots \otimes I.$$

It is then readily verified that

$$a_i^* a_j^* + a_j^* a_i^* = 0; \quad a_i a_j + a_j a_i = 0$$

$$a_i^* a_j + a_j a_i^* = \delta_{ij} I.$$

Furthermore, the combinations entering into T^2 are simply expressed:

$$\sigma_j^- \sigma_{j+1}^- = -a_j a_{j+1} \qquad j = 1,\ldots,2M-1,$$

$$\sigma_j^+ \sigma_{j+1}^+ = a_j^* a_{j+1}^* \qquad j = 1,\ldots,2M-1.$$

On the other hand for $j = 2M$ it is necessary to introduce the auxiliary self adjoint matrix

$$A = \sum_{j=1}^{2M} a_j^* a_j = \sum_{j=1}^{2M} \sigma_j^+ \sigma_j^-.$$

It may be shown that the eigenvalues of A are integers, that A commutes with T^2, that $(-1)^A$ commutes with any monomial of even degree in the a_j's and a_j^*'s and that

$$\sigma_{2M}^- \sigma_1^- = (-1)^A a_{2M} a_1,$$

$$\sigma_{2M}^+ \sigma_1^+ = -(-1)^A a_{2M}^* a_1^*.$$

Hence we have

$$T^2 = \exp\left(-\sum_{j=1}^{2M-1} a_j a_{j+1} + (-1)^A a_{2M} a_1\right).$$

$$\cdot \exp\left(\sum_{j=1}^{2M-1} a_j^* a_{j+1}^* - (-1)^A a_{2M}^* a_1^*\right).$$

Now the eigenvector belonging to the maximum eigenvalue of T^2 is also an eigenvector of A, and it may be shown that the corresponding eigenvalue of A is odd. Thus T^2 reduces to

$$T^2 = \exp\left(-\sum_{j=1}^{2M} a_j a_{j+1}\right) \exp\left(\sum_{j=1}^{2M} a_j^* a_{j+1}^*\right).$$

To diagonalize T^2, we introduce a similar Fourier transformation as before:

$$a_j = \frac{1}{(2Mi)^{1/2}} \sum_{k=k_0}^{2Mk_0} e^{ijk} S_k$$

$$a_j^* = \frac{1}{(-2Mi)^{1/2}} \sum_{k=k_0}^{2Mk_0} e^{-ijk} S_k^*.$$

We have for the anticommutator:

$$S_k S_\ell^* + S_\ell^* S_k = \frac{1}{2M} \sum_{j,j'} e^{-ijk+ij'\ell} (a_j a_{j'}^* + a_{j'}^* a_j) = \delta_{k\ell},$$

and all other anticommutators vanish. Furthermore the exponents of T^2 again become diagonal in the S_k's, resulting in

$$T^2 = \prod_{k=k_0}^{Mk_0} (\Lambda_k) \quad \text{where } \Lambda_k = e^{2S_k S_{-k} \sin k} \, e^{2S_{-k}^* S_k^* \sin k}$$

(we have used the fact that $S_k S_{-k}$ and $S_{-k}^* S_k^*$ do commute with $S_\ell S_{-\ell}$ and $S_{-\ell}^* S_\ell^*$ when $\ell \neq \pm k$). To find the maximum eigenvalue of T^2, it is possible to restrict attention to a special subspace which is invariant under T^2. It is defined by first choosing any vector x_0 which satisfies $S_k x_0 = 0$ for all k, and then constructing the unit vectors

$$e(\delta_1,\ldots,\delta_M) \equiv \prod_{k=k_0}^{Mk_0} (S^*_{-k}S^*_k)^{\delta_k} x_0,$$

$$\delta_k = 0 \quad \text{or} \quad 1.$$

On this basis, it is readily verified that

$$S^*_{-k}S^*_k = I \otimes I \otimes \ldots \otimes I \otimes \begin{pmatrix} 0 & 0 \\ 1 & 0 \end{pmatrix} \otimes I \otimes \ldots \otimes I$$

$$S_kS_{-k} = I \otimes I \otimes \ldots \otimes I \otimes \begin{pmatrix} 0 & 1 \\ 0 & 0 \end{pmatrix} \otimes I \otimes \ldots \otimes I$$

Hence, dropping irrelevant I's,

$$\begin{aligned} \Lambda_k &= \exp\left(2\begin{pmatrix} 0 & 0 \\ 1 & 0 \end{pmatrix}\sin k\right) \exp\left(2\begin{pmatrix} 0 & 1 \\ 0 & 0 \end{pmatrix}\sin k\right) \\ &= \left[I + 2\begin{pmatrix} 0 & 0 \\ 1 & 0 \end{pmatrix}\sin k\right]\left[I + 2\begin{pmatrix} 0 & 1 \\ 0 & 0 \end{pmatrix}\sin k\right] \\ &= \begin{pmatrix} 1 & 2\sin k \\ 2\sin k & 1+4\sin^2 k \end{pmatrix}, \end{aligned}$$

whose maximum eigenvalue is found to be $[\sin k + (1+\sin^2 k)^{1/2}]^2$.

We conclude that the number of configurations is given by

$$\begin{aligned} \mathcal{N}(2M,2N) \simeq [\lambda_{max}(T^2)]^N &= \prod_{k=k_0}^{Mk_0} [\sin k + (1+\sin^2 k)^{1/2}]^{2N} \\ &= \exp 2N \sum_{k=k_0}^{Mk_0} \log(\sin k + (1+\sin^2 k)^{1/2}) \quad \text{or as} \quad M \to \infty \\ &= \exp \frac{2NM}{\pi} \int_0^\pi \log(\sin k + (1+\sin^2 k)^{1/2})dk. \end{aligned}$$

To evaluate the integral, consider

$$I(\gamma) \equiv \int_0^\pi \log[\gamma \sin k + (1+\gamma^2\sin^2 k)^{1/2}]dk$$

Then $I(0) = 0$, while

$$I'(\gamma) = \int_0^{\pi} \frac{\sin k\, dk}{(I+\gamma^2\sin^2 k)^{1/2}} = \int_0^{\pi} \sum_{s=0}^{\infty} \binom{-1/2}{s}\gamma^{2s} \sin^{2s+1} k\, dk$$

$$= \sum_{s=0}^{\infty} \binom{-1/2}{s}\gamma^{2s} B(s-1, \tfrac{1}{2})$$

$$= \sum_{s=0}^{\infty} (-1)^s \frac{\gamma^{2s}}{s+\frac{1}{2}} \quad .$$

It follows that

$$I(1) = \int_0^1 I'(\gamma)d\gamma = 2\sum_{s=0}^{\infty} \frac{(-1)^s}{(2s+1)^2} \equiv 2G,$$

where

$$G = 0.915965594\ldots \text{ Catalan's constant.}$$

Therefore

$$\mathcal{N}(2M,2N) \simeq \exp \frac{G}{\pi}(2M)(2N).$$

Exercises.

1) Find the generating function for the number of non-empty partitions of an integer n, such that no odd part occurs more than once.

2) Prove that if f is a collection of connected labeled graphs, and F the collection of graphs whose components are chosen from f, then the counting functions are related by $F(x) = \exp f(x) - 1$. Define the terms used.

3) Consider a one-dimensional lattice. At each step, the probability of moving one unit to the left is $1/4$, one unit to the right $1/4$, of not moving $1/2$. Find the probability of returning to the initial point after n steps.

3§. The Dimer Problem -- Pfaffian Method.

(M. Fisher and J. Stephenson, Phys. Rev. 132, 1411 (1963) and references therein.)

We now consider a $2M$ by $2N$ square lattice with sharp boundaries and again inquire as to the number $\mathcal{N}(2M,2N)$ of non-overlapping dimer coverings of the lattice. According to the figure oriented counting function discussed in 1§, we have

$$\mathcal{N}(2M,2N) = \text{coef of } \prod_{(p,q)} X^2_{p,q} \text{ in}$$

$$\prod_r (X_{r\ell_1} + X_{r\ell_2} + X_{r\ell_3} + X_{r\ell_4}) \prod_{(p,q)} (1 + X^2_{pq}),$$

where the unordered index pair (p,q) refers to nearest neighbor vertices, and the four-fold sum at vertex r is to be interpreted as containing only three members at an edge or two members at a corner. An alternative interpretation is that in the product over r, each X_{pq} occurs to a power 0, 1, or 2, and we simply want to eliminate the possibility of a unit exponent. One direct way of accomplishing that is by the Gaussian integral method previously discussed in the context of the Master theorem. In this method, one makes use of the identity

$$\int_{-\infty}^{\infty} \begin{pmatrix} 1 \\ x \\ x^2 \end{pmatrix} e^{-x^2/2} \frac{dx}{(2\pi)^{1/2}} = \begin{pmatrix} 1 \\ 0 \\ 1 \end{pmatrix}$$

to rewrite the above expression at once in the form

$$\mathcal{N}(2M,2N) = \int \cdots \int \prod_r \Big(\sum_{i=1}^{4} X_{r\ell_i} \Big) \prod_{(p,q)} \Big(e^{-\frac{1}{2} X^2_{pq}} \frac{dX_{pq}}{(2\pi)^{1/2}} \Big),$$

which while explicit is difficult to evaluate.

A more concise way to eliminate first powers of the X_{pq} is to regard the X_{pq} not as ordinary numbers but as Clifford numbers, members of a non-commutative algebra. The algebraic relations which the Clifford number satisfy are:

$$X_{pq}X_{p'q'} + X_{p'q'}X_{pq} = 2\delta_{(p,q),(p',q')};$$

in other words,

$$X_{pq}X_{p'q'} = -X_{p'q'}X_{pq} \quad \text{if } (p,q) \neq (p',q')$$

while

$$X_{pq}^2 = 1.$$

[It will be noted that the Clifford numbers are closely related to Fermion annihilators and creators, in fact, $X_{pq} = a_{pq} + a^*_{pq} = \begin{pmatrix} 0 & 1 \\ 1 & 0 \end{pmatrix}$, together with anticommutation for different indices.] The crucial fact which we will employ is that: if any matrix representation of the Clifford numbers is introduced, and a normalized trace defined,

$$\mathrm{tr}(X) \equiv \frac{\mathrm{Tr}\,(X)}{\mathrm{Tr}\,(I)}$$

then

$$\mathrm{tr} \prod_d X_d^{\sigma_d} = 1 \quad \text{if} \quad \sigma_d = 0 \quad \text{or} \quad 2 \quad \text{for each pair} \quad (p,q) = d,$$

$$\mathrm{tr} \prod_d X_d^{\sigma_d} = 0 \quad \text{if any} \quad \sigma_d = 1.$$

The first statement follows trivially from the relation $X_d^2 = 1$. To prove the second statement we first eliminate all powers which are 0 or 2, and hence have to show that $\mathrm{tr}\, X_1 X_2 \dots X_m = 0$. Suppose first $m = 2n$; then

$$\mathrm{tr}\, X_1 \dots X_{2n} = -\mathrm{tr}\, X_{2n} X_1 X_2 \dots X_{2n-1}$$

by moving X_{2n} successively to the left using $XY = -YX$ at each stage. But $\mathrm{tr}\, AB = \mathrm{tr}\, BA$; therefore

$$-\mathrm{tr}\, X_{2n} X_1 \dots X_{2n-1} = -\mathrm{tr}\, X_1 X_2 \dots X_{2n-1} X_{2n}$$

which shows that $\mathrm{tr}\, X_1 \dots X_{2n} = 0$. Next suppose $m = 2n-1$; then

$$\begin{aligned} \mathrm{tr}\, X_1 X_2 \dots X_{2n-1} &= \mathrm{tr}\, X_1 \dots X_{2n-1} X_{2n} X_{2n} \\ &= -\mathrm{tr}\, X_{2n} X_1 \dots X_{2n-1} X_{2n} \\ &= -\mathrm{tr}\, X_1 \dots X_{2n-1} X_{2n}^2 \\ &= -\mathrm{tr}\, X_1 \dots X_{2n-1}. \end{aligned}$$

We have thus shown that in the expression

$$\operatorname{tr} \prod_{r} \left(\sum_{i=1}^{4} X_{r\ell_i} \right)$$

every term corresponding to an impossible dimer configuration is absent. Further every term corresponding to an allowed configuration is present and would contribute $+1$ if each pair of X_{pq} which occurred would appear in the product as a pair of adjacent entries: $\ldots X_{pq} X_{pq} \ldots$. However it is clearly impossible for this to be the case in general for every X_{pq}. Thus it is necessary to move a given X_{pq} past a number of other X's in order to place it next to its mate. Since each interchange of X's introduces a minus sign, some of the allowed configurations may appear with weight -1. To avoid this problem we modify the above expression and define

$$\mathcal{N}(2M,2N|C) \equiv \operatorname{tr} \prod_{r} \left(\sum_{i=1}^{4} C_{r\ell_i} X_{r\ell_i} \right).$$

The question is whether we can choose the $C_{r\ell_i}$ such that all non-vanishing terms appear with coefficient $+1$, for if this is achieved we will have shown that $\mathcal{N}(2M,2N|C) = \mathcal{N}(2M,2N)$.

Let us start by finding the sign of a given configuration in $\operatorname{tr} \prod_{r} \left(\sum_{i=1}^{4} X_{r\ell_i} \right)$. We first have to decide upon the ordering of the vertices in the product. We will choose to order the vertices row by row, in lexicographic order i.e. if a vertex is represented by a Cartesian point on the grid then $14 < 15 < 22$. (A zigzag order: to the right on odd rows and to the left on even rows simplifies the form of the $C_{r\ell}$ but complicates everything else.) The process of determining the sign is as follows: we choose a dimer X_{pq} and move it to a position adjacent to the other X_{pq} that must occur. This contributes a sign which is the parity of the number of intervening dimers. We can then remove the term X_{pq}^2 and carry out the same process on some $X_{p'q'}$, afterwards removing $X_{p'q'}^2$ etc. This process is continued until all dimer pairs are exhausted. The trick now is to decide upon a systematic way of carrying out this reduction.

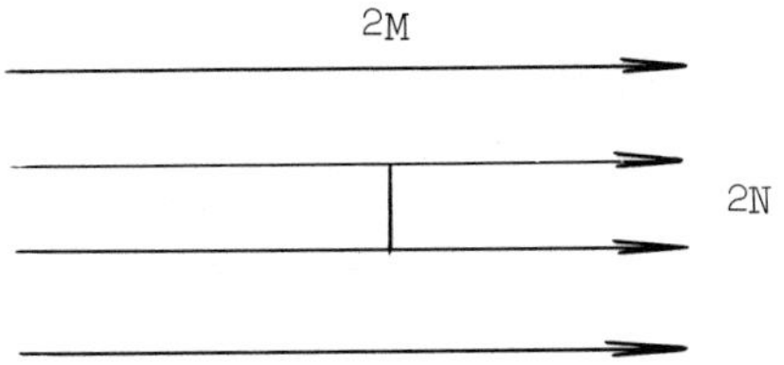

1) Consider a horizontal dimer. It must occur in adjacent terms. Thus its removal contributes a + sign and does not affect the parity of the distance between any two remaining dimers.

2) The bottom row must have an even number of vertical dimers because each horizontal dimer uses up two vertices. Hence there are an even number of vertical dimers between the 2N-1 and 2N rows. It follows easily that there are an even number of vertical dimers connecting any two adjacent rows.

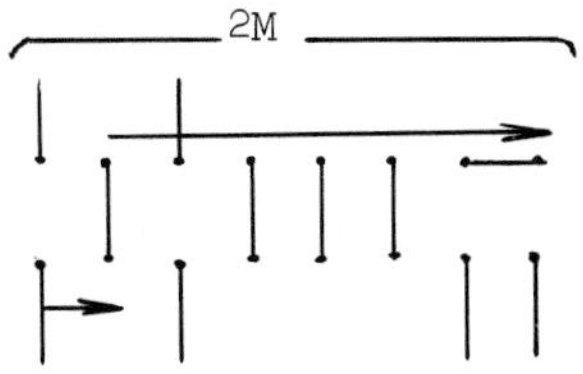

3) Consider a pair of adjacent rows from which no vertical dimers have yet been removed from above or below. Removal of the first vertical dimer involves passing through 2M-1 vertices, then 2M-2 for the second removal etc. Thus each pair of vertical dimers between the two rows inserts a minus sign upon removal.

4) Now consider a pair of adjacent rows from which all vertical dimers (i.e. their vertices) from above or below have been removed, and all horizontal dimers as well. This is equivalent to a short pair of rows totally joined by an even number of vertical dimers and so 3) again applies.

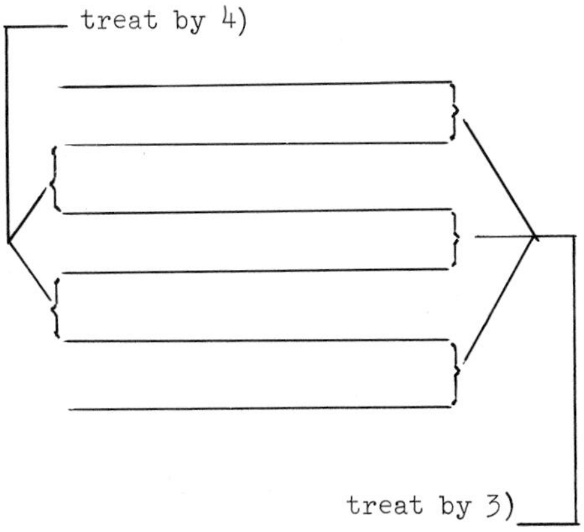

We conclude that a configuration allocates a (-1) to each pair of vertical dimers, so that if we choose

$$C_{r\ell} = \begin{cases} 1 & \text{for horizontal dimer,} \\ i^{1/2} & \text{for vertical dimer,} \\ 0 & \text{otherwise,} \end{cases}$$

then

$$\mathcal{N}(2M,2N) = \mathrm{tr} \prod_{r} \sum_{\ell} C_{r\ell} X_{r\ell}.$$

The evaluation of the trace can now be done recursively. Let us set

$$Q_r \equiv \sum_{\ell} C_{r\ell} X_{r\ell}$$

and observe that

$$\{Q_r Q_s\} \equiv Q_r Q_s + Q_s Q_r = 2 \sum_{\ell} C_{r\ell} C_{s\ell} \equiv 2\, K_{rs}.$$

It follows that for m even

$$\begin{aligned} \mathrm{tr} \prod_{r=1}^{2n} Q_r &= \mathrm{tr}\; Q_1 \ldots Q_{2n} \\ &= \mathrm{tr}\{Q_1 Q_2\} Q_3 \ldots Q_{2n} - \mathrm{tr}\; Q_2 Q_1 Q_3 \ldots Q_{2n} \\ &= \mathrm{tr}\{Q_1 Q_2\} Q_3 \ldots Q_{2n} - \mathrm{tr}\; Q_2 \{Q_1 Q_3\} Q_4 \ldots Q_{2n} \\ &\quad + \mathrm{tr}\; Q_2 Q_3 \{Q_1 Q_4\} Q_5 \ldots Q_{2n} + \ldots - \mathrm{tr}\; Q_2 Q_3 \ldots Q_{2n} Q_1. \end{aligned}$$

Hence

$$\begin{aligned} \mathrm{tr}\; Q_1 Q_2 \ldots Q_n = K_{12}\; &\mathrm{tr}\; Q_3 \ldots Q_{2n} \\ - K_{13}\; &\mathrm{tr}\; Q_2 Q_4 \ldots Q_{2n} \\ + K_{14}\; &\mathrm{tr}\; Q_2 Q_3 Q_5 \ldots Q_{2n} \\ - \ldots\; . \end{aligned}$$

But this is also the recursion relation for the Pfaffian $\mathrm{Pf}(K)$, defined by:

$$Pf(K) \equiv \sum_{\substack{P(2t-1) < P(2t) \\ P(2t-1) < P(2t+1)}} (-1)^P K_{P(1)P(2)} K_{P(3)P(4)} \cdots K_{P(2n-1)P(2n)},$$

as may be verified by expanding by the first term, observing that $P(1) = 1$. Furthermore the two expressions agree when $n = 0$: $tr(1) = 1$, proving that the trace and Pfaffian are identical. The key property of the Pfaffian is that it is also the square root of the corresponding antisymmetric determinant (which will be proved later)

$$Det \; K = [Pf(K)]^2$$

where $K_{sr} = -K_{rs}$ for $r < s$. We conclude that

$$\mathcal{N}(2M, 2N) = [Det \; (K)]^{1/2}$$

where $K_{rs} = \sum_\ell C_{r\ell} C_{s\ell} = -K_{sr}$ for $r < s$, or explicitly in our case

$$K_{rs} = \begin{cases} \pm 1 & \text{if } r \lessgtr s \quad \text{are connected by horizontal dimer,} \\ \pm i & \text{if } r \lessgtr s \quad \text{are connected by vertical dimer,} \\ 0 & \text{otherwise.} \end{cases}$$

Now we are ready to evaluate the determinant. It is best to write the matrix K in compartmental form where each submatrix describes the connections between two rows. Thus

$$K = \begin{pmatrix} A & -iI & & & & \\ iI & A & -iI & & \bigcirc & \\ & iI & A & & & \\ & & & \ddots & & \\ & & & & A & -iI \\ & \bigcirc & & & iI & A \end{pmatrix} \begin{matrix} \text{row 1} \\ \text{row 2} \\ \\ \\ \\ \text{row 2N} \end{matrix}$$

where

$$A = \begin{pmatrix} 0 & 1 & & & \bigcirc \\ -1 & 0 & 1 & & \\ & \ddots & \ddots & \ddots & \\ & & \ddots & \ddots & 1 \\ \bigcirc & & & -1 & 0 \end{pmatrix}, \quad I = \begin{pmatrix} 1 & & & \bigcirc \\ & 1 & & \\ & & \ddots & \\ & & & 1 \\ \bigcirc & & & & 1 \end{pmatrix}$$

are 2M by 2M submatrices. Since all submatrices of K commute, they can be treated as ordinary numbers. Thus if the eigenvalues of

$$A' = \begin{pmatrix} 0 & 1 & & & \bigcirc \\ -1 & 0 & 1 & & \\ & -1 & \ddots & \ddots & \\ & & \ddots & \ddots & 1 \\ \bigcirc & & & -1 & 0 \end{pmatrix} \qquad 2N \text{ by } 2N$$

are $i\lambda_1', \ldots, i\lambda_{2N}'$, then by a similarity transformation, $K = AI' - iA'$ (where $I' = I_{2N \times 2N}$) can be put in the form:

$$K \approx \begin{pmatrix} A+\lambda_1' I & & & \bigcirc \\ & A+\lambda_2' I & & \\ & & \ddots & \\ \bigcirc & & & A+\lambda_{2N}' I \end{pmatrix},$$ where $\approx$ is defined to mean: is similar to.

Hence $\text{Det}(K) = \prod_{s=1}^{2N} \text{Det}(A+\lambda_s' I)$. But if the eigenvalues of A are $i\lambda_1, \ldots, i\lambda_{2M}$ then $\text{Det}(A+\lambda_s' I) = \prod_{t=1}^{2M} (i\lambda_t + \lambda_s')$. It follows that

$$\mathcal{N}(2M, 2N) = [\text{Det}(K)]^{1/2} = \Big[\prod_{s=1}^{2N} \prod_{t=1}^{2M} (\lambda_s' + i\lambda_t)\Big]^{1/2}.$$

To find the eigenvalues of A, we observe that if

$$D_{2M} \equiv \text{Det}\,(A-i\lambda) = \text{Det}\begin{pmatrix} -i\lambda & 1 & & \\ -1 & -i\lambda & \ddots & \\ & \ddots & \ddots & 1 \\ & & -1 & -i\lambda \end{pmatrix}_{2M \times 2M}$$

then expanding by the first row,

$$D_p = -i\lambda D_{p-1} + D_{p-2}$$

with boundary conditions:

$$D_1 = -i\lambda, \quad D_0 = 1.$$

The difference equation for D_p has characteristic equation

$$x^2 + i\lambda x - 1 = 0,$$

and if we set $\lambda = 2 \sin \theta$ the roots are:

$$x_1 = e^{-i\theta}, \quad x_2 = -e^{i\theta}.$$

Therefore $D_p = ax_1^p + bx_2^p$, or using the boundary conditions

$$D_p = \frac{1}{2 \cos \theta} [(e^{-i\theta})^{p+1} - (-e^{i\theta})^{p+1}].$$

The eigenvalues are given by $D_{2M} = 0$, or $\cos (2M+1)\theta = 0$, $\cos \theta \neq 0$. Hence

$$\lambda_s = 2 \sin \theta_s \quad \text{where} \quad \theta_s = \frac{s - \frac{1}{2}}{2M+1} \pi, \quad s = 1-M, \ldots, M,$$

and similarly

$$\lambda'_t = 2 \sin \theta'_t \quad \text{where} \quad \theta_t = \frac{t - \frac{1}{2}}{2N+1} \pi, \quad t = 1-N, \ldots, N.$$

We conclude that

$$\mathscr{N}(2M,2N) = [2^{4MN} \prod_{s=1-M}^{M} \prod_{t=1-N}^{N} (\sin \frac{s-\frac{1}{2}}{2M+1}\pi + i \sin \frac{t-\frac{1}{2}}{2N+1}\pi)]^{1/2}$$

which can be reduced to

$$2^{2MN} \prod_{s=1}^{M} \prod_{t=1}^{N} [\sin^2 \frac{s-\frac{1}{2}}{2M+1}\pi + \sin^2 \frac{t-\frac{1}{2}}{2N+1}\pi].$$

As special cases it can be shown that

$$\mathscr{N}(4,4) = 2^2\, 3^2$$

$$\mathscr{N}(8,8) = 2^4 (901)^2 \qquad \text{and}$$

$$\mathscr{N}(2M,2N) \xrightarrow[\substack{M \to \infty \\ N \to \infty}]{} e^{\frac{4MN}{\pi}G}$$

where G is Catalan's constant.

Finally we return to the important Pfaffian expansion theorem which we used in the form: If K is an antisymmetric matrix of even order, the determinant of K is denoted by $|K|$, its ij cofactor by $|K|_{ij}$, and higher cofactors analogously, then

$$|K|^{1/2} = \sum_{i=2}^{2n} K_{1i}(-1)^{i-1}|K|_{\substack{11\\ii}}^{1/2} .$$

It suffices to prove the square of this relationship, namely,

$$|K| = \sum_{i,j} K_{1i}K_{1j}(-1)^{i+j}\, |K|_{\substack{11\\ii}}^{1/2} |K|_{\substack{11\\jj}}^{1/2} .$$

Now according to the Jacobi Identity (see Aitken: Determinants and Matrices, p.99),

$$|K'|_{ii}|K'|_{jj} - |K'|_{ij}|K'|_{ji} = |K'|\,|K'|_{\substack{ii\\jj}}$$

for any K'; choosing K' as K with first row and first column removed, and noting that an odd antisymmetric determinant vanishes, we have

$$|K|_{\substack{11\\ii}} |K|_{\substack{11\\jj}} = |K|_{\substack{11\\ij}} |K|_{\substack{11\\ji}} .$$

Taking the square root and determining the sign by comparing products of anti-diagonal terms this becomes

$$|K|^{1/2}_{\substack{11\\ii}} |K|^{1/2}_{\substack{11\\jj}} = (-1)^{i+j} |K|_{\substack{11\\ij}} .$$

Hence we have to prove that

$$|K| = \sum_{i,j} K_{1i} K_{1j} |K|_{\substack{11\\ij}} .$$

But according to the Cauchy expansion by first row and first column

$$|K| = |K|_{11} - \sum_{i,j} K_{1i} K_{j1} |K|_{\substack{11\\ij}} .$$

Since $|K|_{11} = 0$ and $K_{j1} = -K_{1j}$ the proof is complete.

<u>Exercises.</u>

1) Find the probability of returning to the origin after n steps for a nearest neighbor isotropic random walk on a triangular lattice.

2) Consider a ballot in which at each step A receives a block of μ votes or B receives one vote. Suppose that after $n = a+b$ trials, A has received μa votes, B has received b votes, <u>and that $b \geq \mu a$</u>. Find the probability of t ties during the balloting.

3) Evaluate the number of dimer coverings on an 8 by 8 lattice.

4§. <u>The Dimer Problem - First Permanent Method.</u>

The transfer matrix approach avoids topological problems completely, while the Pfaffian technique does make use of a substantial topological property. By going as quickly as possible to a clear topological question, some of the mysteries of the Pfaffian approach can be circumvented and a relatively brief derivation achieved.

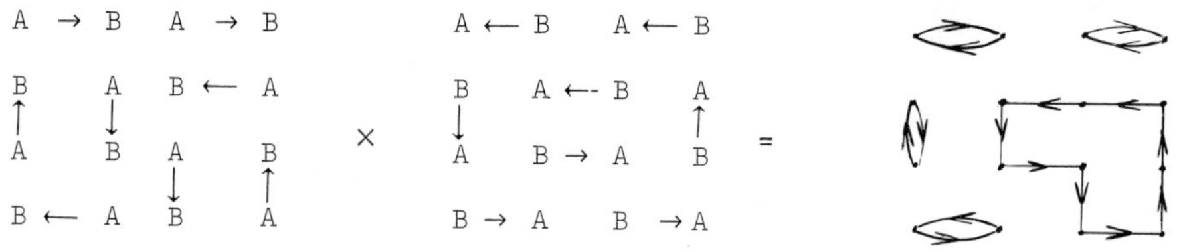

I II III

Suppose we decompose our grid into A and B sublattices. Each dimer must connect an A vertex and a B vertex. In a dimer configuration, say I, let us give each dimer an $A \to B$ orientation. In an independent dimer configuration, say II, we give each dimer a $B \to A$ orientation. Then in the superposition, denoted by III = I × II, every vertex has one incoming and one outgoing arrow. Thus I × II is completely covered by non-intersecting oriented closed curves. Further, any such covering III separates uniquely into a type I and a type II dimer configuration. We conclude that the number of possible oriented loop coverings $\mathcal{N}_c$ is the square of the number of possible dimer configurations:

$$\mathcal{N}_c(2M,2N) = [\mathcal{N}(2M,2N)]^2.$$

Our immediate topological problem is then to find the number of coverings by oriented closed curves. This takes on a different aspect in permanent language. A covering in fact is identical with a transition in which each vertex -- the tail of the arrow -- is carried to a different vertex -- the head of the arrow -- such that the second set is identical with the first. We conclude that a covering is simply a permutation of the set of vertices, with the restriction that the only transitions permitted are to a horizontal or vertical neighbor. The number of such permutations is of course a permanent:

$$\mathcal{N}_c = \operatorname{Per} \begin{pmatrix} B & I & & \\ I & B & \ddots & \\ & \ddots & \ddots & I \\ & & I & B \end{pmatrix} \begin{matrix} \text{row } 1 \\ \\ \\ \text{row } 2N \end{matrix} \equiv \operatorname{Per}(L)$$

row 1 .... row 2N

where

$$B = \begin{pmatrix} 0 & 1 & & & \\ 1 & 0 & 1 & & \\ & \ddots & \ddots & \ddots & \\ & & \ddots & \ddots & 1 \\ & & & 1 & 0 \end{pmatrix}, \quad I = \begin{pmatrix} 1 & & \bigcirc \\ & \ddots & \\ \bigcirc & & 1 \end{pmatrix}$$

are the $2M \times 2M$ matrices of a row to a row.

The trick of evaluation is now to alter signs such that the permanent becomes equal to a determinant, which can then be evaluated. In other words, we choose suitable multipliers $\bar{\ell}_{rs} \propto \ell_{rs}$ such that $\mathrm{Per}\,(L) = \mathrm{Det}\,\bar{L}$ or

$$\sum_P \ell_{1P(1)} \cdots \ell_{4MN,P(4MN)} = \sum_P (-1)^P\, \bar{\ell}_{1P(1)} \cdots \bar{\ell}_{4MN,P(4MN)},$$

because $\bar{\ell}_{1P(1)} \cdots \bar{\ell}_{4MN,P(4MN)} = (-1)^P \ell_{1P(1)} \cdots \ell_{4MN,P(4MN)}$ for every permutation P. It is not obvious that this can be done, although the fact that many terms vanish identically is a big help. To find out whether suitable $\bar{\ell}_{rs}$ can be determined, we must find the parity of the permutation corresponding to any given type III configuration. This is most easily obtained by dividing the permutations into cycles, noting that a cycle is precisely a closed loop of vertices on the lattice. Further, each loop has an even number of vertices, so that each loop contributes -1 to the parity (since an odd number of transpositions are required for an even cycle). It follows that the parity of the configuration is just that of the number of closed curves it contains.

Let us characterize a closed curve R by its loop number $L(R)$, the number of horizontal bond $h(R)$, vertical bonds $v(R)$, and interior points $I(R)$. We further

R $\qquad\qquad$ $\bar{R}$

introduce a condensation operation $R \to \bar{R}$ which effectively covers the vertices of the loop by dimers, as shown: the odd links are doubled and the even links dropped (or vice versa, it will not matter). We then claim that

<u>Theorem.</u> $L(R) - I(R) + \frac{1}{4}(h(R) - v(R)) \equiv L(\bar{R}) - I(\bar{R}) + \frac{1}{4}(h(\bar{R}) - v(\bar{R}))[\text{mod } 2]$.

Of course, for a single loop, $L(R) = 1$ and $I(\bar{R}) = 0$.

It is trivial to verify this theorem for the basic curves ,  and . Any other curve may be built up by adding squares one at a time, each one having one or two sides in common with the previous curve. Thus we have to show that the theorem is unchanged under either of these basic operations. We indicate this schematically:

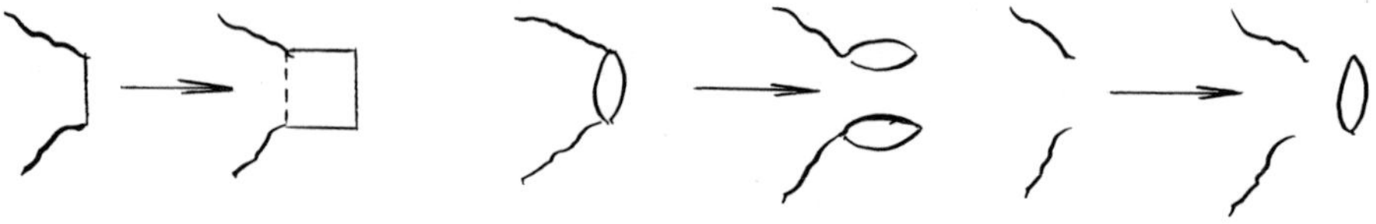

R	$\bar{R}$	$\bar{R}'$
$\Delta L = 0$	$\Delta L = 1$	$\Delta L = 1$
$\Delta I = 0$	$\Delta I = 0$	$\Delta I = 0$
$\Delta h = 2$	$\Delta h = 4$	$\Delta h = 0$
$\Delta v = 0$	$\Delta v = -2$	$\Delta v = 2$
$\Delta(L-I+\frac{1}{4}(h-v)) = \frac{1}{2}$	$\frac{5}{2}$	$\frac{1}{2}$

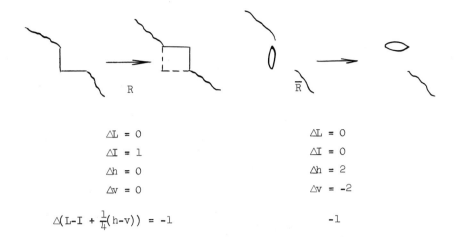

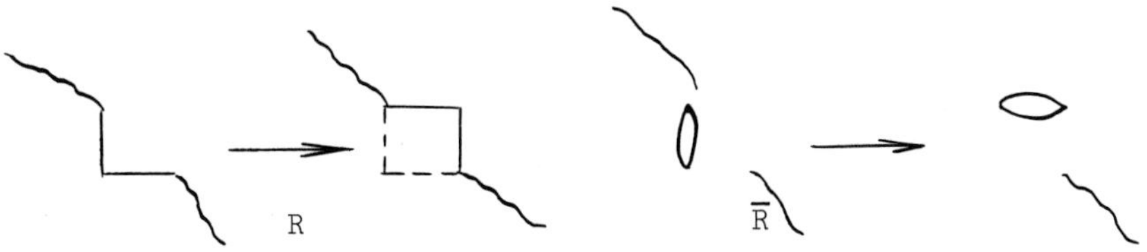

$\Delta L = 0$ $\qquad\qquad$ $\Delta L = 0$

$\Delta I = 1$ $\qquad\qquad$ $\Delta I = 0$

$\Delta h = 0$ $\qquad\qquad$ $\Delta h = 2$

$\Delta v = 0$ $\qquad\qquad$ $\Delta v = -2$

$\Delta(L-I + \frac{1}{4}(h-v)) = -1$ $\qquad\qquad$ -1

Thus the theorem is established inductively. Let us rewrite it in somewhat simpler form, first by remarking that $L(\overline{R}) = \frac{1}{2}(h(\overline{R}) + v(\overline{R}))$, and then by subtracting 1/4 of the identity $h(R) + v(R) = h(\overline{R}) + v(\overline{R})$ from both sides. We obtain for a single loop:

<u>Corollary.</u> $L(R) - I(R) \equiv \frac{1}{2} v(R) + \frac{1}{2} h(\overline{R}) \pmod 2$.

Obviously $I(R) \equiv 0 \pmod 2$ for any loop belonging to a complete covering because the interior must be covered by dimers. Summing over all loops we then have

$$\sum L(R) \equiv \frac{1}{2} \sum v(R) + \frac{1}{2} \sum h(\overline{R}) \pmod 2,$$

but $\frac{1}{2} \sum h(\overline{R})$ is the number of horizontal dimers (pairs of horizontal links) in the covering and we have proved that it is even. It follows that the parity of a covering is given by

$$(-1)^P = (-1)^{\sum L(R)} = (-1)^{\frac{1}{2} \sum v(R)} = i^{\sum v(R)}.$$

Thus we have shown that if $\overline{\ell}_{rs} = i\, \ell_{rs}$ for every vertical link and $\overline{\ell}_{rs} = \ell_{rs}$ for every horizontal link, then $\mathscr{N}_c = \mathrm{Det}\,(\overline{L})$. Written out in detail,

$$\bar{L} = \begin{pmatrix} B & iI & & & \bigcirc \\ iI & B & \ddots & & \\ & \ddots & \ddots & \ddots & \\ & & \ddots & \ddots & iI \\ \bigcirc & & & iI & B \end{pmatrix}$$

and proceeding as in §3 we can then show that

$$\mathcal{N}_c = \prod_s \prod_t (\lambda_s + i\lambda_t'),$$

where λ_s and λ_t' are the eigenvalues of $B_{2M \times 2M}$ and $B_{2N \times 2N}$ respectively. These eigenvalues can be obtained directly as before, or more simply by observing that

$$DBD^{-1} = iA$$

where

$$D = \begin{pmatrix} 1 & & & & & & \\ & i & & & & & \\ & & -1 & & & & \\ & & & -i & & & \\ & & & & \cdot & & \\ & & & & & \cdot & \\ & & & & & & \cdot \end{pmatrix}$$

and A is the basic matrix of §3. Thus the eigenvalues of B and iA and the result of §3 is reproduced.

5§. <u>The Dimer Problem -- Second Permanent Method.</u>

The representation of a pair of dimer configurations by a set of simple closed paths and hence by a permanent is valuable to indicate the genesis of the Pfaffian or square root of a determinant. However, this is an artificial if elegant method. We would hope more directly to have a single dimer configuration correspond to a single permanent -- a standard allowed permutation problem -- which can then be evaluated.

An allowed permutation is a transition in which a set of vertices goes over to the same set of vertices. Now a dimer can be regarded as causing such a transition *if* the vertices on the B sublattice are put in 1-1 correspondence with the vertices on the A sublattice each in lexicographic order. Here we simply draw an arrow from

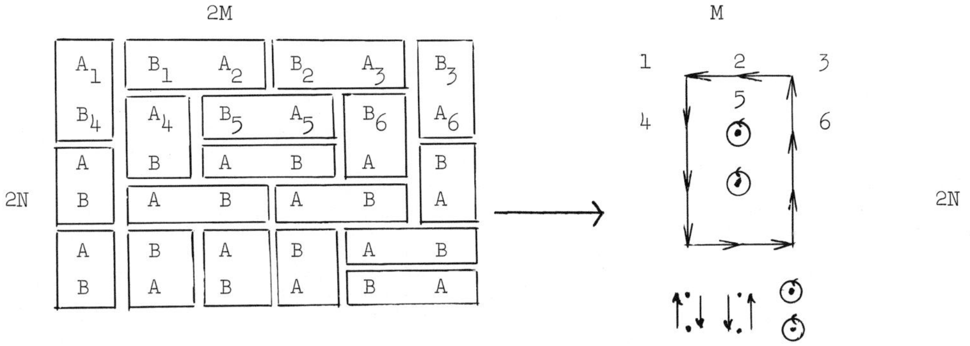

an A vertex to the appropriate B vertex, each expressed as an integer between 1 and 2MN. There results again a set of closed loops, not completely general, for one has a necessary and sufficient condition that an arrow can go left only on an odd row and go right only on an even row. This condition has the following consequences:

i) There exist loops covering only a simple vertex.

ii) A two-vertex loop must be vertical.

iii) A many-vertex loop must have an even number of interior points.

[In fact any connected sequence of interior points along a vertical line must be bounded by one left and one right arrow, since otherwise they could not all be interior. Thus each such connected sequence consists of an even number of points.]

possible impossible

The permanent which counts all permutations is again obtained by piecing together the $M \times M$ submatrices which represent the possible transitions between one row and another. Here one has to distinguish between odd and even rows. One has:

$$\text{odd} \to \text{odd}: \qquad C = \begin{pmatrix} 1 & & & & \bigcirc \\ 1 & 1 & & & \\ & 1 & 1 & & \\ & & \ddots & \ddots & \\ \bigcirc & & & 1 & 1 \end{pmatrix} \qquad M \times M$$

$$\text{odd} \to \text{even}: \quad I, \qquad \text{even} \to \text{odd}: \quad I$$

$$\text{even} \to \text{even}: \qquad C^T = \begin{pmatrix} 1 & 1 & & \bigcirc \\ & 1 & \ddots & \\ & & \ddots & 1 \\ \bigcirc & & & 1 \end{pmatrix}$$

in terms of which $\mathcal{N}(2M, 2N) = \mathrm{Per}(K)$ where

$$K = \begin{pmatrix} C & I & & & & \bigcirc \\ I & C^T & I & & & \\ & I & C & I & & \\ & & I & C^T & \ddots & \\ & & & \ddots & \ddots & I \\ \bigcirc & & & & I & C^T \end{pmatrix}, \text{ a } 2N \times 2N \text{ compartmentalized matrix.}$$

We want to write $\mathrm{Per}(K) = \mathrm{Det}(\overline{K})$ and so again require the parity of a loop configuration. If the one-vertex loop (of <u>odd</u> length) is called trivial, this parity is then given by the number of non-trivial loops. Since the interior of each loop is even, then according to the Corollary of 4§ with the roles of horizontal and vertical links interchanged, the total number of non-trivial loops is given by:

$$\sum L(R) \equiv \frac{1}{2} \sum h(R) + \frac{1}{2} \sum v(\overline{R}) \pmod 2.$$

Let us denote by v_{2j} the total number of vertical dimers to row $2j$ in the dimer covering which is equivalent to the condensed set $\{\overline{R}\}$, and by s_{2j} the total number of trivial vertices on row $2j$. Then clearly

$$v_{2j} + s_{2j} \equiv M \pmod 2$$

so that

$$\frac{1}{2} \sum v(\overline{R}) = \sum_{j=1}^{N} v_{2j} \equiv NM + \sum_{j=1}^{N} s_{2j}.$$

We conclude that

$$(-1)^P = (-1)^{\sum L(R)} = (-1)^{MN} (-1)^{\sum_{j=1}^{N} s_{2j}} {}_i\sum h(R).$$

It follows that

$$\mathscr{N}(2M, 2N) = (-1)^{MN} \operatorname{Det}(\overline{K}) \quad \text{where } \overline{K} = \begin{pmatrix} \overline{c} & I & & & \bigcirc \\ I & -\overline{c}^{+} & I & & \\ & I & \overline{c} & \ddots & \\ & & \ddots & \ddots & I \\ \bigcirc & & & I & -\overline{c}^{+} \end{pmatrix}$$

and

$$\overline{c} = \begin{pmatrix} 1 & & & \bigcirc \\ i & 1 & & \\ & \ddots & \ddots & \\ \bigcirc & & i & 1 \end{pmatrix} \qquad \overline{c}^{+} = \begin{pmatrix} 1 & -i & & \bigcirc \\ & 1 & -i & \\ & & \ddots & \ddots \\ & & & -i \\ \bigcirc & & & 1 \end{pmatrix}$$

Since $\overline{K}$ does not consist of commuting submatrices, we require a preliminary reduction. Multiply every even column by $\overline{c}$ on the right; then subtract from it the preceeding odd column. We obtain:

$$\begin{pmatrix}
\bar{c} & 0 & 0 & 0 & & & \\
I & -\bar{c}^{+}\bar{c}-2I & I & -I & & \bigcirc & \\
0 & 0 & \bar{c} & 0 & & & \\
0 & -I & I & -\bar{c}^{+}\bar{c}-2I & & & \\
 & & & 0 & \ddots & & \\
 & & & -I & & \ddots & -I \\
 & \bigcirc & & & & \ddots & 0 \\
 & & & & & & -\bar{c}^{+}\bar{c}-2I
\end{pmatrix}_{2N \times 2N}$$

Here every odd row has only one entry, namely $\bar{c}$. Expanding out by these elements and reversing signs MN times we have: $\mathcal{N}(2M,2N) = \operatorname{Det}(\bar{\bar{K}})$, where

$$\bar{\bar{K}} = \begin{pmatrix}
2I+\bar{c}^{+}\bar{c} & I & & \bigcirc \\
I & 2I+\bar{c}^{+}\bar{c} & \ddots & \\
 & \ddots & 2I+\bar{c}^{+}\bar{c} & I \\
\bigcirc & & I & I+\bar{c}^{+}\bar{c}
\end{pmatrix}_{N \times N}$$

and

$$\bar{c}^{+}\bar{c} = \begin{pmatrix}
2 & -i & & & \bigcirc \\
i & 2 & \ddots & & \\
 & i & \ddots & \ddots & \\
 & & \ddots & 2 & -i \\
\bigcirc & & & i & 1
\end{pmatrix}_{M \times M}$$

Consider the matrix

$$J_p = \begin{pmatrix}
2 & 1 & & & \bigcirc \\
1 & 2 & \ddots & & \\
 & 1 & \ddots & 2 & 1 \\
\bigcirc & & \ddots & 1 & 1
\end{pmatrix}_{p \times p}$$

But the discriminant of a cubic equation is given by

$$\mathrm{Disc}(\lambda^3+a_1\lambda^2+a_2\lambda+a_3 = 0) = a_1^2a_2^2 - 4a_1^3a_3 - 27a_3^2 + 18a_1a_2a_3.$$

Thus

$$\mathscr{N}(6,6) = 2^3\,\frac{41209}{49} = 2^3\cdot 29^2.$$

D. Counting Patterns on Two Dimensional Lattices.

1§. The Ice Problem -- Introduction.

The water molecule, H_2O, has two hydrogen atoms equivalent and at an obtuse angle. When water molecules come together, they can be bound by a quite strong hydrogen bond, the hydrogen still belonging to one oxygen but holding onto a foreign oxygen at a greater distance.

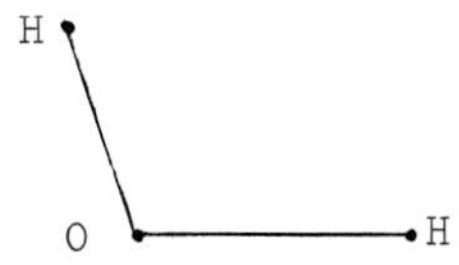

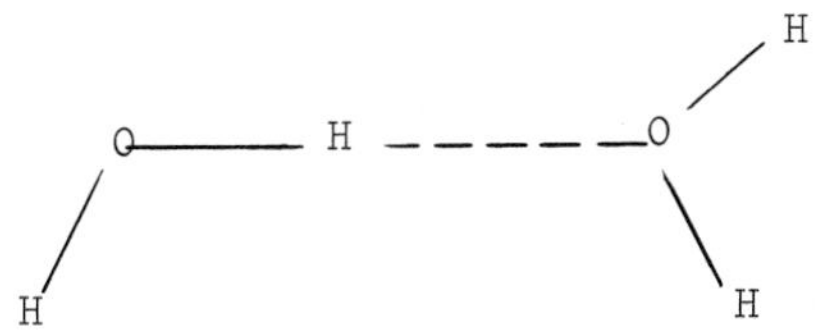

In fluid H_2O, as many H-bonds will form as possible, consistent with the thermal motion of the O's; ionic bonds are possible but rare, since $(OH)^-$ and $(H_3O)^+$ exist in water only to the extent of 1 in 10^7. For low temperature ice, we then expect a rigid oxygen structure with maximum H bonding: each O has two H's which it shares with two other O's. Thus there are 4n H bonds for n O's, which means that each O has precisely two nearby H's and two further away. At normal pressure, it is also known that the oxygen crystal structure is hexagonal wurtzite, built from a body-centered cubic lattice, as shown, with each O surrounded by a tetrahedron of nearest neighbors.

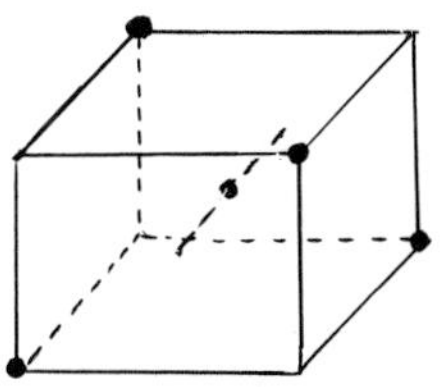

There are many ways of taking an oxygen lattice and orienting H bonds to keep integral water molecules, i.e. to have exactly two nearby H's. This so-called ice condition (equivalent to charge neutrality: each O is neutralized by two H's) is schematically represented by directing an arrow into the close O-H space, and demanding two arrows in, two arrows out per vertex (Oxygen site).

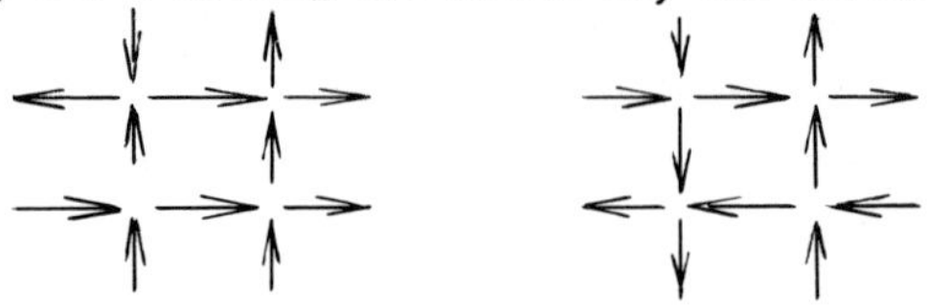

This degeneracy in H-bond placement alone exists at temperature 0° and is responsible for the residual entropy (violation of the third law of thermodynamics) of ice, which may in fact be measured by starting with the gaseous entropy and going down to zero temperature via the specific heat: $S(0) = S(T) - \int_0^T \frac{1}{T} C_v(T)dT$.

Microscopically, the entropy per site, $S \equiv S(0)$, is given by

$$S = k \frac{1}{n} \log \mathcal{N}(n),$$

where $\mathcal{N}(n)$ is the number of possible configurations for n sites. If we write:

$$\mathcal{N}(n) \equiv [W(n)]^n,$$

then in the asymptotic limit $n \to \infty$,

$$S = k \log W,$$

where $W = \lim_{n \to \infty} W(n)$.

Pauling's estimate of W is independent of the lattice structure, obtained only from the configuration number, 4, and the ice condition. He observes that there are $2n$ bonds, each with two possible orientations, for a total of 2^{2n} possible patterns. However, at a vertex, only 6 of the possible 16 bond configurations

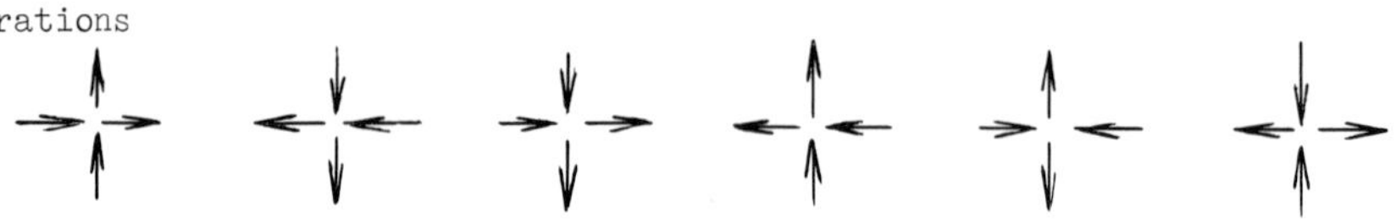

are allowed by the ice condition. Some of the prohibited configurations may be prohibited more than once (inclusion-exclusion principle is required for a complete counting by this technique) and so we conclude that

$$\mathcal{N} \geq 2^{2n}\left(\frac{6}{16}\right)^n$$

or

$$W \geq 1.5.$$

The exact result for real ice is $W = 1.507\ldots$ A poor upper bound for the two dimensional version of ice, which we will discuss more fully in a moment, is given by the fact that the complete configuration is certainly determined by the arrows on all vertical columns together with those on a single horizontal row, so that

$$\mathcal{N} \leq 2^n . 2^{n^{1/2}}$$

or

$$W \leq 2.$$

For better estimates and exact answers, one must specify the form of the lattice. Eschewing reality (no exact solution exists) we shall look at square ice, something like a projection of real ice onto the plane: here, arrows are placed on a square lattice to satisfy the ice condition. One can improve on Pauling's estimate by considering clusters of vertices. For example, for a square of four vertices, there are 2^{12} a priori possibilities for the 12 associated links. Let us designate each of the central four links by + if counter clockwise, - if clockwise (thus the configuration shown is ++++). It is easily verified that the following number of allowed possibilities exists:

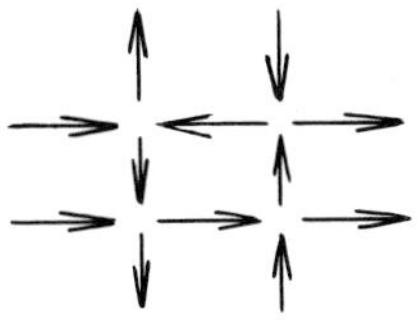

+ + + +	16	+ - + -	2
+ + + -	16	+ - - -	16
+ + - -	16	- - - -	16

where we have not distinguished between rotated configurations e.g. + + + - equals + + - + . We conclude that

$$W \geq \left(\frac{82}{2^{12}}\right)^{1/4} 2^2 = 1.505.$$

This technique converges slowly to the exact 1.540.

A less specialized and more effective procedure is due primarily to Nagle (Jour. Math. Phys. Aug. 1966). Here one attempts a direct counting of possible vertex configurations. Suppose the six possibilities at the i-th vertex are labeled by $\xi_i = 1,\ldots,6$. For neighboring vertices i,j, we then define

$$a(\xi_i,\xi_j) = \begin{cases} +1 & \text{if the configurations at } i,j \text{ are consistent} \\ -1 & \text{otherwise.} \end{cases}$$

(Consistency demands only that the single arrow connecting the vertices leaves one vertex and enters the other.) Hence

$$\mathcal{N} = \sum_{\{\xi_i\}} \prod_{i<j} \tfrac{1}{2}(1 + a(\xi_i,\xi_j)),$$

the sum being over n vertices and the product over $2n$ links. Pulling out a factor of 6 from each vertex we have

$$\mathcal{N} = (\tfrac{3}{2})^n \sum_{\{\xi_i\}} 6^{-n} \prod_{i<j} (1 + a(\xi_i,\xi_j)),$$

or expanding out the product,

$$\mathcal{N} = (\tfrac{3}{2})^n \sum_{\{\xi_i\}} 6^{-n}[1 + \sum_{i<j} a(\xi_i,\xi_j) + \sum_{i<j} \sum_{\substack{k<\ell \\ (i,j)\neq(k,\ell)}} a(\xi_i,\xi_j)a(\xi_k,\xi_\ell) + \ldots].$$

Now if we fix the configuration of one vertex, a neighboring vertex has 3 consistent configurations and 3 inconsistent ones. Thus

$$\sum_{\xi_i=1}^{6} a(\xi_i,\xi_j) = 1 + 1 + 1 - 1 - 1 - 1 = 0,$$

and in the same way we can show that

$$\sum_{\xi_i=1}^{6} a(\xi_i,\xi_j)a(\xi_i,\xi_k)a(\xi_i,\xi_\ell) = 0.$$

This means that in the above expansion of the product, the only terms which will not vanish on summation are those for which each vertex occurs zero, two, or four times. Expressed diagrammatically, each surviving term will be represented by a set of links such that each vertex which is present connects either 2 or 4 links.

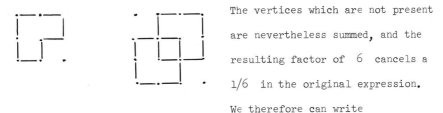

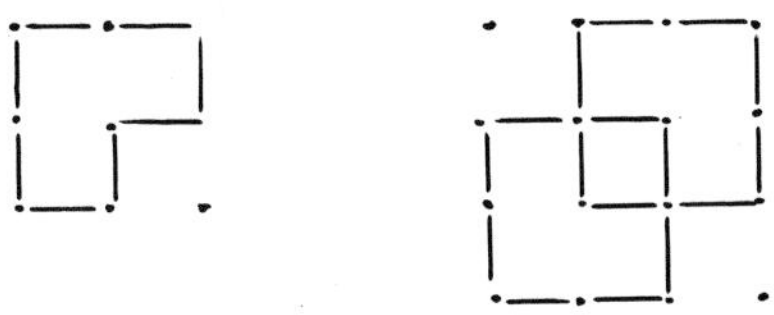

The vertices which are not present are nevertheless summed, and the resulting factor of 6 cancels a 1/6 in the original expression. We therefore can write

$$\mathcal{N} = \left(\tfrac{3}{2}\right)^n \sum_{\text{graphs } G} \frac{1}{6^G} \sum_{\{\xi_i:\ i \in G\}} \prod_{\{i<j:\ i,j \in G\}} a(\xi_i, \xi_j),$$

or more briefly

$$\mathcal{N} = \left(\tfrac{3}{2}\right)^n \sum_G W(G).$$

Suppose now that the $\{C_\alpha\}$ $\alpha = 1,\ldots,\Omega$ are the full set of inequivalent rooted connected graphs, where two graphs are regarded as equivalent if they differ only by a rigid translation on the lattice or by the choice of the root. We can, therefore, place one such graph (including the empty graph) at each vertex provided that no two subgraphs overlap:

$$\mathcal{N} = \left(\tfrac{3}{2}\right)^n \sum_{\substack{\{\alpha_j=1\} \\ \text{non-overlap}}}^{\Omega} \prod_{j=1}^{n} W(C_{\alpha_j}).$$

In order to remove the restriction of non-overlapping, we use the inclusion-exclusion principle, first writing down the sum without restrictions, then subtracting the sum with a simple restriction i.e. at least one overlap, adding the sum of two restrictions -- at least two overlaps -- etc. Explicitly, let $\{C_\beta^{(1)}\}$ be the set of graphs with exactly one overlap, connected by virtue of the overlap, let $\{C_\gamma^{(2)}\}$ be the set of graphs similarly connected by exactly two overlaps,... The above expression may clearly be written as:

$$\mathcal{N} = \left(\tfrac{3}{2}\right)^n \Big\{ \sum_{\{\alpha_j=1\}}^{\Omega} \prod_{j=1}^{n} W(C_{\alpha_j}) - \sum_{\substack{\{\alpha_j=1\} \\ j \neq i}}^{\Omega} \sum_{\{\beta_i=1\}}^{\Omega_1} W(C_{\beta_i}^{(1)}) \prod_{j \neq i} W(C_{\alpha_j}) + \ldots \Big\}$$

where the next term has either a double overlapping graph at one vertex or single overlapping graphs at two vertices, ... It is not hard to show that this is just the multinomial expansion of

$$\mathcal{N} = \left(\tfrac{3}{2}\right)^n \Big\{\sum_\alpha W(C_\alpha) - \sum_\beta W(C_\beta^{(1)}) + \sum_\gamma W(C_\gamma^{(2)}) - \dots\Big\}^n$$

and we conclude that (see Domb. Adv. Phys. 9, 149, (1960))

$$W(n) = \frac{3}{2}\Big\{\sum_\alpha W(C_\alpha) - \sum_\beta W(C_\beta^{(1)}) + \sum_\gamma W(C_\gamma^{(2)}) - \dots\Big\}.$$

As $n \to \infty$ each of the sums in $W(n)$ becomes a convergent infinite series and thus we obtain W. It may be shown (see Nagle) that a graph of n links and C crossovers has an associated weight of 3^{2C-n}. The connected graphs with values 4, 6 and 8 for $n-2C$ and their multiplicities are given as follows:

$n - 2C = 4$ 

$n - 2C = 6$ 2:  2: 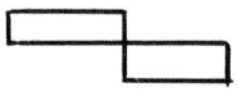

$n - 2C = 8$ 2:  4: 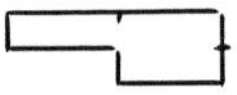 8: 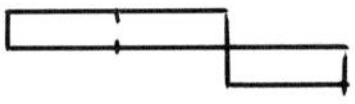

4: 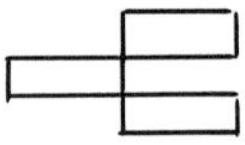 2: 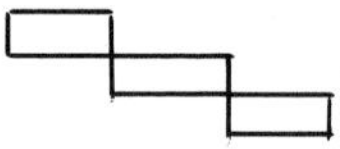 1: 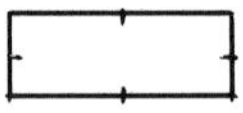

1: 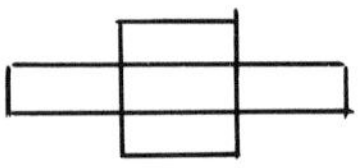

whereas the lowest overlapping graphs, with $n_1+n_2 - 2(c_1+c_2) = 8$ are

2: 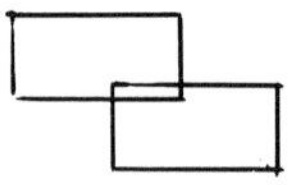 and 2: 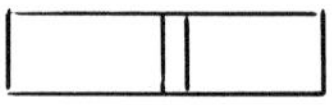

Hence

$$W = \frac{3}{2}\left(1 + \frac{1}{3^4} + \frac{4}{3^6} + \frac{22 - 4}{3^8} + \dots\right)$$

$$= 1.531 + \dots \longrightarrow 1.540.$$

2§. Square Ice -- The Transfer Matrix Method.

(E. Lieb, Phys. Rev. Letters 18, 692 (1967).)

We now seek the exact solution, with periodic boundary conditions on an N by N lattice. Again we use the arrow notation, with the ice condition at each vertex. A row configuration may be specified by the orientation of its upper vertical arrows, +1 for up, -1 for down -- or, as a unit vector in a 2^N dimensional space, by the Cartesian product of $\binom{1}{0}$ for each up arrow, $\binom{0}{1}$ for each down arrow. To show that it is sufficient to focus upon the upper vertical arrows, we observe that a change in upper arrow direction between

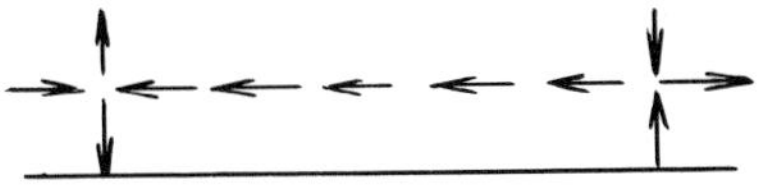

adjacent rows uniquely determines the horizontal arrows until the next change, and demands that the next change will be in the opposite direction i.e. for an up → down change the next change to left or right must be from down to up. There are also two singular cases, that in which no changes occur; here it is clear that the horizontal arrows can be either all to the left or all to the right. We conclude that the transfer matrix which gives the weight of the transition from a row configuration u_{j-1} to a row configuration u_j is specified by

$$T(u_j, u_{j-1}) = \begin{cases} 1 & \text{if arrow changes alternate in direction} \\ 2 & \text{if there are no changes: } u_j = u_{j-1} \\ 0 & \text{otherwise.} \end{cases}$$

The relevance of the transfer matrix now follows as in the corresponding discussion for the dimer problem. Starting with the bottom row, u_N, a unit vector in a 2^N dimensional space, the possible succeeding rows are determined by

$$u_N \xrightarrow{T} u_{N-1} \xrightarrow{T} u_{N-2} \xrightarrow{T} \ldots,$$

but the zero-th and N-th rows are to be identical, so that

$$\mathcal{N} = \sum_{u_N} (u_N, T^N u_N) = \mathrm{Tr}\,(T^N).$$

But if $\lambda_{max} > 0$ is the maximum eigenvalue of T, and if T is diagonalizable, clearly

$$\mathcal{N} = \mathrm{Tr}\,(T^N) = \sum_{\alpha=1}^{2^N} \lambda_\alpha^N ;$$

hence

$$\lambda_{max}^N \leq \mathcal{N} \leq 2^N \lambda_{max}^N$$

or

$$\lambda_{max}^{1/N} \leq \mathcal{N}^{1/N^2} \leq 2^{1/N} \lambda_{max}^{1/N} .$$

Since

$$W = \lim_{N \to \infty} \mathcal{N}^{1/N^2}$$

it follows that

$$W = \lim_{N \to \infty} \lambda_{max}^{1/N} .$$

It will be somewhat more convenient to adopt the lattice gas notation for a row configuration, specifying only the locations of the up arrows, i.e. $u(x_1, x_2, \ldots, x_s)$ has up arrows at $x_1, x_2, \ldots, x_s$, down arrows at the remaining N-s locations. Suppose then that the j-th row has configuration $u(x_1, \ldots, x_s)$, the (j-1)-th $u(y_1, \ldots, y_s)$. Then, as we proceed from left to right, there are two possibilites. The first up arrow may belong to the bottom row, i.e. y_1. If an up arrow never occurs at the same location on both rows, then the condition of alternate changes requires $y_1 < x_1 < y_2 < x_2 < \ldots < x_s$. If a pair of up arrows can occur then we can have $x_i = y_i$, but the remaining inequalities must be maintained. Hence for this first class of transitions we have

$$T_1 \colon \; y_1 \leq x_1 \leq y_2 \leq x_2 \leq \ldots \leq x_s .$$

The remaining possibility is that x_1 occurs first, and then by the same argument we have

$$T_2: x_1 \leq y_1 \leq x_2 \leq y_2 \leq \cdots \leq y_s.$$

(Note that due to periodic boundary conditions, x_s must be followed by y_1 in T_1, y_s by x_1 in T_2, which guarantees that the number of up arrows is the same (s) for all rows.) We conclude that the unit vector $u(x_1,\ldots,x_s)$ can make transitions according to the rule:

$$Tu(x_1,\ldots,x_s) = \sum_{y_1=1}^{x_1} \sum_{y_2=x_1}^{x_2} \cdots \sum_{y_s=x_{s-1}}^{x_s} u(y_1,\ldots,y_s)$$

$$+ \sum_{y_1=x_1}^{x_2} \sum_{y_2=x_2}^{x_3} \cdots \sum_{y_s=x_s}^{N} u(y_1,\ldots,y_s),$$

with the convention that

$$u(y_1,\ldots,y_s) \equiv 0 \quad \text{if} \quad y_i = y_{i+1} \quad \text{for some} \quad i.$$

The operations involved in T are successive indefinite sums (like indefinite integrals) so that one would imagine that T can be reduced considerably by successive differencing operations. Indeed, if we define the translation operator for the j-th arrow by $E_j f(x_j) = f(x_j+1)$, then

$$(E_s-1) \sum_{y_s=x_{s-1}}^{x_s} u(y_s) = \sum_{y_s=x_{s-1}}^{x_s+1} u(y_s) - \sum_{y_s=x_{s-1}}^{x_s} u(y_s) = u(x_s+1) = E_s u(x_s),$$

$$(E_{s-1}-1) \sum_{y_{s-1}=x_{s-2}}^{x_{s-1}} u(y_{s-1},x_s+1) = u(x_{s-1}+1,x_s+1)$$

$$= E_{s-1}E_s u(x_{s-1},x_s), \text{ etc.,}$$

whereas

$$(E_1-1) \sum_{y_1=x_1}^{x_2} u(y_1) = \sum_{y_1=x_1+1}^{x_2} u(y_1) - \sum_{y_1=x_1}^{x_2} u(y_1) = -u(x_1),$$

$$(E_2-1) \sum_{y_2=x_2}^{x_3} - u(x_1,y_2) = \sum_{y_2=x_2+1}^{x_3} - u(x_1,y_2) + \sum_{y_2=x_2}^{x_3} u(x_1,y_2)$$

$$= u(x_1,x_2,), \text{ etc.}$$

Thus

$$\prod_{j=1}^{s}(E_j-1)T\ u(x_1,\ldots,x_s) = \left(\prod_{j=1}^{s}E_j + (-1)^s\right)u(x_1,\ldots,x_s).$$

It is crucial to observe however that the above holds only if $x_i - x_{i-1} > 1$ for every $i = 1,\ldots,s$, since otherwise the E_j need not commute: e.g.

$$E_1E_2\ u(x_1,x_2)\Big|_{x_2=x_1+1} = u(x_1+1,x_1+2)$$

whereas

$$E_2E_1\ u(x_1,x_2)\Big|_{x_2=x_1+1} = 0.$$

In the allowed domain the eigenvalue equation $Tu = \lambda u$ then becomes

$$\left(\prod_{j=1}^{s}E_j + (-1)^s\right)\ u(x_1,\ldots,x_s) = \lambda\prod_{j=1}^{s}(E_j-1)\ u(x_1,\ldots,x_s).$$

For orientation let us suppose that the above equation is required to be valid over the complete region $x_1 < x_2 < x_3 < \ldots$; this will, of course, result in a lower bound to the maximum eigenvalue. The boundary condition in this case is that $u = 0$ whenever $x_i = x_{i+1}$ for some i. It will be convenient to extend the eigenfunction over the whole space by setting

$$u(x_{P(1)},x_{P(2)},\ldots,x_{P(s)}) \equiv (-1)^P u(x_1,x_2,\ldots,x_s)$$

for any permutation P. The consistency of this definition is equivalent to the boundary condition since the only points in common between two regions defined by different permutations occur when two x's coincide; the definition requires that $u = 0$ at such a point, which is precisely the boundary condition. Thus we may restrict attention to antisymmetric solutions with periodicity as the only remaining boundary condition. The eigenfunction equation is a symmetric linear difference equation with constant coefficients. Hence its basic solution is given by

$$e^{ik_1x_1}\ e^{ik_2x_2}\ \ldots\ e^{ik_sx_s}$$

where periodicity requires that $k_j = K_j \frac{2\pi}{N}$, K_j integer, and the general anti-symmetric solution is

$$u(x_1,\dots,x_s) = \sum_{k_1,\dots,k_s} c(k_1,\dots,k_s)(-1)^P e^{i\sum_{j=1}^{s} k_j x_{P(j)}}$$

Clearly, $k_j \neq k_\ell$ for $j \neq \ell$ and we may restrict our attention to increasing sequences $(k_1,\dots,k_s)$. Finally, since in the original situation, a function not depending on the location of one arrow could not possibly have corresponded to the maximum configuration number, we shall assume that $k_j \neq 0$ for $j = 1,\dots,s$.

Now substituting the basic solution into the eigenvalue equation, we see that the k_j's are restricted by

$$(e^{ik_1+\dots+ik_s} + (-1)^s)u(x_1,\dots,x_s) = \lambda \prod_{j=1}^{s}(e^{ik_j} - 1)u(x_1,\dots,x_s)$$

and hence that the eigenvalue corresponding to $u(x_1,\dots,x_s)$ is

$$\lambda = \frac{e^{i\sum_{j=1}^{s} k_j} + (-1)^s}{\prod_{j=1}^{s}(e^{ik_j} - 1)} = \frac{e^{\frac{i}{2}\sum_{j=1}^{s} k_j} + (-1)^s e^{-\frac{i}{2}\sum_{j=1}^{s} k_j}}{\prod_{j=1}^{s}(e^{\frac{i}{2}k_j} - e^{-\frac{i}{2}k_j})}$$

or

$$\lambda = \begin{cases} \dfrac{2\cos\frac{1}{2}\sum_{j=1}^{s} k_j}{i^s \prod_1^s 2\sin\frac{1}{2}k_j} & \text{if } s \text{ is even;} \\[2ex] \dfrac{2i\sin\frac{1}{2}\sum_{j=1}^{s} k_j}{i^s \prod_1^s 2\sin\frac{1}{2}k_j} & \text{if } s \text{ is odd.} \end{cases}$$

Forgetting the numerator for the moment, the maximum λ is obtained by choosing as many factors $2\sin\frac{1}{2}k_j$ to be less than unity as possible, and hence choosing all $k_j \neq 0$ in the range $|k_j/2| < \pi/6$. Suppose s is even; then $k_j = j\frac{2\pi}{N}$, $j = \pm 1, \pm 2, \dots, \pm s/2$, and $s = \frac{N}{3}$ (or $\frac{N}{3} - 1$), so that

$$\lambda = \frac{2}{\prod_{j=1}^{[\frac{N}{6}]} (2 \sin \frac{j\pi}{N})^2} .$$

The same value is obtained when s is odd. Explicitly we then have

$$\log W = \frac{1}{N} \log \lambda$$

$$= - \frac{2}{N} \sum_{j=1}^{[N/6]} \log 2 \sin \frac{j\pi}{N}$$

$$= - \frac{2}{\pi} \int_0^{\pi/6} \log 2 \sin \theta \, d\theta$$

$$= \frac{2}{\pi} \int_0^{\pi/6} (\cos 2\theta + \frac{1}{2} \cos 4\theta + \frac{1}{3} \cos 6\theta + \ldots) d\theta$$

$$= 0.32.$$

Hence $W = e^{0.32} = 1.38$ which is indeed a lower bound for square ice. Furthermore we observe that in this case the maximum eigenfunction is unique, i.e., non-degenerate.

3§. Square Ice -- Exact Solution.

We now proceed to the determination of the maximum eigenvalue of the transfer matrix with the correct boundary conditions. It will be valuable to start by obtaining two characteristic properties which must be satisfied by the maximum eigenvalue and eigenvector.

First, we shall fix the number s of up arrows on each row which is responsible for the maximum eigenvalue. Let us denote the maximum eigenvalue for s up arrows on N sites by $\lambda_s(N)$ and the value of s which maximizes $\lambda_s(N)$ by $s(N)$. Then clearly

$$W = \lim_{N \to \infty} [\lambda_{s(N)}(N)]^{1/N}.$$

Furthermore, by reversing up and down arrows we have as well

$$W = \lim_{N \to \infty} [\lambda_{N-s(N)}(N)]^{1/N}.$$

Now as $N \to \infty$ there will be no distinction between periodic boundary conditions and fixed arrow boundary conditions. We shall write $\bar{\lambda}$ to refer to the situation in which the arrows on the left and right edges are required to point up. Hence,

$$(*) \qquad W = \lim_{N\to\infty} [\bar{\lambda}_{s(N)}(N)]^{1/N}, \quad W = \lim_{N\to\infty} [\bar{\lambda}_{N-s(N)}(N)]^{1/N}.$$

But if we juxtapose an s arrow configuration on a 2N by N lattice with fixed boundary, and an N-s arrow configuration on a similar lattice, we obtain a valid N arrow configuration on a 2N by 2N lattice with fixed boundary. Thus

$$\bar{\mathcal{N}}_{s(N)}(2N,N)\bar{\mathcal{N}}_{N-s(N)}(2N,N) \le \bar{\mathcal{N}}_{N}(2N,2N),$$

which implies that

$$\lim_{N\to\infty} [\bar{\lambda}_{s(N)}(N)]^{1/2N}[\bar{\lambda}_{N-s(N)}(N)]^{1/2N} \le \lim_{N\to\infty} [\bar{\lambda}_{N}(2N)]^{1/2N}.$$

Since $\lim_{N\to\infty} [\bar{\lambda}_{N}(2N)]^{1/2N} \le W$, whereas from (*)

$$\lim_{N\to\infty} [\bar{\lambda}_{s(N)}(N)]^{1/2N}[\bar{\lambda}_{N-s(N)}(N)]^{1/2N} = W,$$

we conclude that

$$\lim_{N\to\infty} [\lambda_{N}(2N)]^{1/2N} = \lim_{N\to\infty} [\bar{\lambda}_{N}(2N)]^{1/2N} = W,$$

and hence that

$$\lim_{N\to\infty} \frac{s(2N)}{2N} = \frac{1}{2}.$$

Next we shall show that the maximal eigenvector of the transfer matrix T is unique and has all components positive. To do this requires a theorem, due to Perron and Frobenius.

Theorem. If a real square matrix $T = (T_{\alpha\beta})$ satisfies the conditions

i) $T_{\alpha\beta} \ge 0$ for all α and all β;

ii) For fixed β, $T_{\alpha\beta} > 0$ for at least one α;

iii) For every pair (α,β) there exists an n such that $(T^n)_{\alpha\beta} > 0$;

and if λ is the eigenvalue of maximum absolute value, then

i) λ is real and positive,

ii) λ has only one eigenvector, whose components are positive real.

<u>Proof.</u> Consider the transformation

$$v_\alpha \to T_\alpha(v) \equiv \frac{\sum_\beta T_{\alpha\beta} v_\beta}{\sum_\alpha \sum_\beta T_{\alpha\beta} v_\beta}$$

for vectors satisfying: $v_\alpha \geq 0$, $\sum_\alpha v_\alpha = 1$. Since at least one v_α is greater than zero, then by condition ii) we must have $\sum_\alpha \sum_\beta T_{\alpha\beta} v_\beta > 0$. Hence the transformation is continuous and takes a closed connected space into itself. By the Brouwer fixed point theorem the transformation has at least one fixed point: there exists a vector u with $u_\alpha \geq 0$ such that:

$$u_\alpha = \frac{\sum_\beta T_{\alpha\beta} u_\beta}{\sum_\alpha \sum_\beta T_{\alpha\beta} u_\beta}$$

Therefore i) $Tu = \lambda u$ where $\lambda = \sum_\alpha \sum_\beta T_{\alpha\beta} u_\beta > 0$. Suppose now that v is a different eigenvector (v is not a scalar multiple of u); then

$$\sum_\beta T_{\alpha\beta} v_\beta = \lambda' v_\alpha .$$

The proof of ii) then proceeds in four stages.

a) Suppose that all the v_α are real and non-negative. Let $\rho = \operatorname*{Min}_\beta \frac{u_\beta}{v_\beta}$. Then $u_\beta - \rho v_\beta \geq 0$ and

$$u_\gamma - \rho v_\gamma > 0 \quad \text{for some } \gamma, \quad \text{and}$$

$$u_\alpha - \rho v_\alpha = 0 \quad \text{for some } \alpha.$$

It follows that

$$0 = \sum_\beta T_{\alpha\beta}(u_\beta - \rho v_\beta) = \lambda u_\alpha - \lambda' v_\alpha \rho$$

$$= (\lambda - \lambda') u_\alpha,$$

so that $\lambda' < \lambda$.

b) In general, define $\rho = \operatorname{Min}_\beta \frac{u_\beta}{|v_\beta|}$, and suppose that for some γ, $u_\gamma - \rho|v_\gamma| > 0$.

If α is chosen such that $u_\alpha - \rho|v_\alpha| = 0$, then by condition iii) we can select an integer n for which $(T^n)_{\alpha\gamma} > 0$. It follows that:

$$0 < \sum_\beta (T^n)_{\alpha\beta}(u_\beta - \rho|v_\beta|).$$

But

$$\sum_\beta (T^n)_{\alpha\beta} u_\beta = \lambda^n u_\alpha$$

and

$$\sum_\beta (T^n)_{\alpha\beta}|v_\beta| \geq |\lambda'|^n |v_\alpha|.$$

Hence

$$0 < \lambda^n u_\alpha - |\lambda'|^n |v_\alpha| \rho = (\lambda^n - |\lambda'|^n) u_\alpha$$

and we conclude that

$$|\lambda'| < \lambda.$$

c) Suppose now that $u_\gamma - \rho|v_\gamma| = 0$ for all γ. Then proceeding as in b) we have instead

$$0 \leq (\lambda^n - |\lambda'|^n) u_\alpha \quad \text{for all } \alpha.$$

Thus $|\lambda'| \leq \lambda$ and must only eliminate the case $|\lambda'| = \lambda$. But if $|\lambda'| = \lambda$ and $|v_\gamma| = \frac{1}{\rho} u_\gamma$, then we have both

$$\sum_\beta T_{\alpha\beta} v_\beta = \lambda' v_\alpha$$

$$\sum_\beta T_{\alpha\beta}|v_\beta| = |\lambda'||v_\alpha|.$$

Therefore there exists some scalar θ such that

$$T_{\alpha\beta} v_\beta \frac{1}{\theta}$$

is non-negative real for all α, β, and consequently $\frac{1}{\theta} v_\beta$ is a non-negative real eigenvector. We have disposed of this case in a).

d) We have seen that the eigenvector u belonging to the maximum absolute eigenvalue λ is unique and may be written so that its components are real non-negative:

$$u_\beta \geq 0.$$

Now choose any α. For any fixed γ for which $u_\gamma > 0$, choose n such that $(T^n)_{\alpha\gamma} > 0$. We then have

$$u_\alpha = \frac{1}{\lambda^n} \sum_\gamma (T^n)_{\alpha\gamma} u_\gamma > 0.$$

Therefore the components of u are in fact positive. Q.E.D.

The transfer matrix for the square ice model is easily shown to satisfy the conditions of this theorem. We recall that in the ice problem, for a specified number s of up arrows per row, the unit vectors were given by all $\binom{N}{s}$ possible configurations of s tuples on the N sites. The general vector u is then a linear combination of the $\binom{N}{s}$ unit vectors, and T is a square matrix on this $\binom{N}{s}$ dimensional space of mixed configurations. The elements of T in the basis of s up arrow states have been shown to have the values 2, 1 or 0, depending on whether they are diagonal, connect consistent adjacent configurations, or inconsistent ones. Condition i) of the Theorem is trivial, since $T_{\alpha\beta} = \begin{cases} 0 \\ 1; \\ 2 \end{cases}$ condition ii) likewise since $T_{\alpha\alpha} = 2$. To prove the irreducibility of T, i.e. condition iii), we observe that if the configuration α is placed at row 1, configuration β at row $N+1$, a set of consistent intermediate configurations can be obtained. We do this by connecting corresponding arrows by straight lines and, proceeding from α to β, change the location of an arrow on a row each time it passes an integer location. Hence $(T^n)_{\alpha\beta} > 0$ for all α and β. It follows then from the theorem that at the specified value $s = N/2$, there is only one maximal eigenvector, and its components can be chosen as positive. Furthermore since T is a symmetric matrix, its eigenvectors can be chosen as real and orthogonal. But then only one eigenvector can have all components positive, and this uniquely specifies the maximal eigenvector.

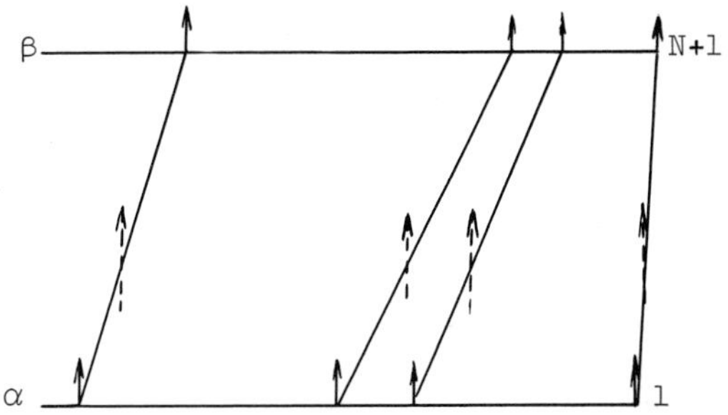

We now return to the exact solution of the maximal eigenvector of the transfer matrix. Since the difference equation is still satisfied in the region in which $|x_i - x_j| > 1$, we can still write

$$u(x_1,\ldots,x_s) = \sum_{k_1,\ldots,k_s} c(k_1,\ldots,k_s) e^{i \sum_{j=1}^{s} k_j x_j} ,$$

over those sets of $k_1,\ldots,k_s$ for which

$$\lambda = \frac{e^{i \sum_{j=1}^{s} k_j} + (-1)^s}{\prod_{j=1}^{s} (e^{ik_j} - 1)}$$

However, since the domain of validity is not connected, we cannot complete u to an antisymmetric function. All we can do is note that when $k_1,\ldots,k_s$ is an allowed set, then so is any permutation $k_{P(1)},\ldots,k_{P(s)}$. Thus as basic solution we can take the Bethe ansatz

$$u(x_1,\ldots,x_s) = \sum_P a(P) e^{i \sum_{j=1}^{s} k_{P(j)} x_j} ;$$

it is of course possible that a linear combination of such expressions for different sets of consistent $k_1,\ldots,k_s$ will be required, but this turns out to be unnecessary. The boundary conditions are of two types, periodic and internal

($x_i = x_{i-1} + 1$). It will be convenient first to examine the effect of translation invariance: if each location x_i is translated to the right by 1, another maximal eigenvector results, which by normalization and uniqueness must be identical with the original one. Hence $u(x_1+1,\ldots,x_s+1) = u(x_1,\ldots,x_s)$ which implies at once that

$$e^{i\sum_{j=1}^{s} k_j} = 1 \quad \text{or} \quad \sum_{j=1}^{s} k_j = 2\pi n, \; n \; \text{integer}.$$

Now consider the effect of periodic boundary conditions. By combining with translation invariance, we clearly have $u(x_1,\ldots,x_s) = u(1,x_2-x_1+1,\ldots, x_s-x_1+1) = u(x_2-x_1,\ldots,x_s-x_1,N) = u(x_2,\ldots,x_s, N+x_1)$. Thus if we define a cyclic permutation c such that

$$c(i) = i+1, \quad c(s) = 1$$

we have

$$\sum_P a(P) e^{i\sum_{j=1}^{s} k_{P(j)} x_j} = \sum_P a(P) e^{i(\sum_{j=1}^{s} k_{P(j)} x_{c(j)} + Nk_{P(s)})}$$

$$= \sum_P a(P) e^{i(\sum_{j=1}^{s} k_{Pc^{-1}(j)} x_j + Nk_{P(s)})}$$

$$= \sum_Q a(Qc) e^{i(\sum_{j=1}^{s} k_{Q(j)} x_j + Nk_{Q(1)})} .$$

We conclude that

$$a(Q) = a(Qc) e^{iNk_{Q(1)}}.$$

The internal boundary conditions are substantially more difficult to apply. Perhaps the simplest approach is to solve a comparison system which requires the same boundary conditions. For this purpose, it is helpful to reintroduce a spin matrix notation, in which the base vector at the j-th site, $\binom{1}{0}$ for up spin (direction of arrow) $\binom{0}{1}$ for down spin is acted upon by

$$S_j^- = \begin{pmatrix} 0 & 0 \\ 1 & 0 \end{pmatrix}, \quad S_j^+ = \begin{pmatrix} 0 & 1 \\ 0 & 0 \end{pmatrix}, \quad S_j^z = \begin{pmatrix} 1 & 0 \\ 0 & -1 \end{pmatrix}.$$

One then sees that the transfer matrix may be written as

$$T = A + A^{+}$$

where

$$A = I + \sum_{i<j} S_i^- S_j^+ + \sum_{i<j<k<\ell} S_i^- S_j^+ S_k^- S_\ell^+ + \dots$$

$$A^{+} = I + \sum_{i<j} S_i^+ S_j^- + \sum_{i<j<k<\ell} S_i^+ S_j^- S_k^+ S_\ell^- + \dots ,$$

since the operation $S_i^- S_j^+$ moves an up spin from the i-th site to the j-th site, leaving other sites unchanged, etc. Now consider the auxiliary matrix with periodic boundary conditions,

$$H = \sum_{i=1}^{N} (S_i^- S_{i+1}^+ + S_i^+ S_{i+1}^- + \frac{1}{2} S_i^z S_{i+1}^z).$$

[In the notation $S^+ = \frac{1}{\sqrt{2}}(S^x + iS^y)$, $S^- = \frac{1}{\sqrt{2}}(S^x - iS^y)$, this becomes

$$H = \sum_{i=1}^{N} (S_i^x S_{i+1}^x + S_i^y S_{i+1}^y + \frac{1}{2} S_i^z S_{i+1}^z),$$

and is familiar to the anisotropic Heisenberg model.] It may be shown after some tedious computation that the operators T and H commute, and it follows that we may choose a simultaneous complete set of eigenvectors for T and H. In particular, the unique eigenvector of H with all components positive will be the maximal eigenvector of T.

If we decompose H as

$$H = \sum_{i=1}^{N} (B_{i,i+1} + D_{i,i+1})$$

where

$$B_{i,i+1} = S_i^- S_{i+1}^+ + S_i^+ S_{i+1}^-, \quad D_{i,i+1} = \frac{1}{2} S_i^z S_{i+1}^z,$$

then with respect to the pair $(i,i+1)$ we have

$$B_{i,i+1} \begin{cases} \uparrow\uparrow = 0 \\ \downarrow\downarrow = 0 \\ \uparrow\downarrow = \downarrow\uparrow \\ \downarrow\uparrow = \uparrow\downarrow \end{cases} \qquad D_{i,i+1} \begin{cases} \uparrow\uparrow = \frac{1}{2}\uparrow\uparrow \\ \downarrow\downarrow = \frac{1}{2}\downarrow\downarrow \\ \uparrow\downarrow = -\frac{1}{2}\uparrow\downarrow \\ \downarrow\uparrow = -\frac{1}{2}\downarrow\uparrow \end{cases} .$$

In our previous notation in which only the positions of up arrows are specified, B moves an up arrow one site to the left or right if there is room for it, or otherwise yields zero; under the same circumstances, D yields $+ 1/2$ or $- 1/2$, respectively. Thus

$$\begin{aligned} Hu(x_1,\dots,x_s) = {} & \sum_{j=1}^{N} (1-\delta_{x_{j+1},x_j+1}) u(x_1,\dots,x_{j-1},x_j+1,x_{j+1},\dots,x_s) \\ & + \sum_{j=1}^{N} (1-\delta_{x_{j-1},x_j-1}) u(x_1,\dots,x_{j-1},x_j-1,x_{j+1},\dots,x_s) \\ & + \sum_{j=1}^{N} (-\tfrac{1}{2} + \delta_{x_{j+1},x_j+1}) u(x_1,\dots,x_s). \end{aligned}$$

If the δ-functions are disregarded, the eigenvector equation $Hu = \lambda u$ becomes a linear difference equation, and we may make the Bethe ansatz

$$u(x_1,\dots,x_s) = \sum a(P) e^{i \sum_j k_{P(j)} x_j}$$

over the full domain $x_1 \le x_2 \le \dots \le x_s$. But the coefficient of δ_{x_{j+1},x_j+1} is:

$$\begin{aligned} & u(x_1,\dots,x_{j-1},x_j,x_j+1,x_{j+2},\dots,x_s) - \\ & u(x_1,\dots,x_{j-1},x_j+1,x_j+1,x_{j+2},\dots,x_s) - \\ & u(x_1,\dots,x_{j-1},x_j,x_j,x_{j+2},\dots,x_s), \end{aligned}$$

so our boundary condition is that this expression must vanish for all j.

Inserting the boundary conditions into the ansatz, we get

$$\sum_P a(P) e^{i \sum_{t \neq j, j+1} k_{P(t)} x_t} [e^{i(k_{P(j)} + k_{P(j+1)})(x_j+1)} + e^{i(k_{P(j)} + k_{P(j+1)}) x_j} - e^{i(k_{P(j)} x_j + k_{P(j+1)}(x_j+1))}]$$

$$\equiv 0.$$

If Q is that permutation which is identical with P, except on the $j, j+1$ positions:

$$P(t) = Q(t) \quad \text{for} \quad t \neq j, \; j+1$$

$$k_{P(j)} = k_{Q(j+1)} \equiv k$$

$$k_{P(j+1)} = k_{Q(j)} \equiv q,$$

then since on defining $x_{j+1} = x_j$, we have

$$\sum_{t=1}^{N} k_{P(t)} x_t = \sum_{t=1}^{N} k_{Q(t)} x_t,$$

it follows that

$$\frac{1}{2} \sum_P e^{i \sum_{t=1}^{N} k_{P(t)} x_t} \{a(P)[e^{i(k+q)} + 1 - e^{iq}] + a(Q)[e^{i(k+q)} + 1 - e^{ik}]\} = 0.$$

Therefore

$$\frac{a(P)}{a(Q)} = - \frac{e^{i(k+q)} + 1 - e^{ik}}{e^{i(k+q)} + 1 - e^{iq}}$$

whenever P and Q differ only by a $j \leftrightarrow j+1$ transposition. Since any transposition can be constructed from neighboring transpositions: $(1t) = (12)(23)(34)\ldots(t-1,t)\ldots(34)(23)(12)$, any cycle from transpositions, and any permutation from cycles, we have thus determined all $a(P)$ to within normalization. To normalize we may set $a(I) = 1$.

Finally, to find the k_j, we must make use of the periodic boundary condition $a(Q) = a(Qc) e^{iNk_{Q(1)}}$. Choosing $Q = I$ and noting that $c = (12)(23)\ldots(s-1,s)$ we have

$$e^{-iNk_1} = a((12)(23)\ldots(s-1,s))$$

$$= -a((12)(23)\ldots(s-2,s-1))\,\frac{1+e^{ik_s}e^{ik_1}-e^{ik_s}}{1+e^{ik_s}e^{ik_1}-e^{ik_1}}$$

$$= \ldots$$

$$= (-1)^s \prod_{j=2}^{s} \left(\frac{1+e^{ik_j}e^{ik_1}-e^{ik_j}}{1+e^{ik_j}e^{ik_1}-e^{ik_1}}\right).$$

Hence for s even

$$e^{-iNk_1} = \prod_{j=1}^{s} \left(\frac{1+e^{ik_j}e^{ik_1}-e^{ik_j}}{1+e^{ik_j}e^{ik_1}-e^{ik_1}}\right),$$

which must remain valid under a cyclic rotation, and we conclude that

$$e^{-iNk_p} = \prod_{j=1}^{s} \left(\frac{1+e^{ik_j}e^{ik_p}-e^{ik_j}}{1+e^{ik_j}e^{ik_p}-e^{ik_p}}\right),$$

$$p = -\tfrac{s}{2},\ldots,-1,1,\ldots,\tfrac{s}{2}.$$

This set of equations can be solved in principle for $k_{-s/2},\ldots,k_{s/2}$, but the solution is not unique. We would like that solution for which $u(x_1,\ldots,x_s)$ remains positive throughout. As we saw when analyzing the approximate difference equation, the maximum eigenvalue for T will be obtained by making the k_p as small as possible, without any two of them being equal, and not allowing $k = 0$. This will be achieved (subject to explicit verification later of the positivity of $u(x_1,\ldots,x_s)$) by taking the logarithm of the above equation and selecting the branches of minimum absolute value:

$$iNk_p = 2\pi ip + \sum_{j=1}^{s} \log\left(1+e^{ik_p}e^{ik_j}-e^{ik_p}\right)$$

$$- \sum_{j=1}^{s} \log\left(1+e^{ik_p}e^{ik_j}-e^{ik_j}\right).$$

In order to solve for the k_p, we shall go at once to the limit $N \to \infty$.

In this limit, the k_j which must lie between $-\pi$ and π become infinitely dense and we can speak only of the distribution $\rho(k)$:

$$\frac{1}{s}\sum_{j=-s/2}^{s/2} F(k_j) = \int F(k)\rho(k)dk$$

where

$$\int \rho(k)dk = 1.$$

Therefore

$$k_p = \frac{2\pi}{N} p + i\frac{s}{N}\int \log(1 + e^{iq}e^{ik_p} - e^{iq})\rho(q)dq$$

$$- i\frac{s}{N}\int \log(1 + e^{iq}e^{ik_p} - e^{ik_p})\rho(q)dq.$$

But since $s\rho(k)\Delta k = \Delta p$ we have

$$\rho(k) = \frac{1}{s}\frac{\partial p}{\partial k}$$

for any k in the range $k_{-s/2}$ to $k_{s/2}$ (which need not extend from $-\pi$ to π), and so by differentiating with respect to k we get

$$1 = 2\pi\frac{s}{N}\rho(k) - \frac{s}{N}\int \frac{\rho(q)e^{i(k+q)}}{1 + e^{i(k+q)} - e^{iq}} dq$$

$$+ \frac{s}{N}\int \frac{\rho(q)e^{ik}(e^{iq}-1)}{1 + e^{i(k+q)} - e^{ik}} dq.$$

Since this equation is only valid within the range of k, it is important to know what this range is. It is in fact determined by the condition that the solution $\rho(k)$ satisfy the normalization condition $\int \rho(k)dk = 1$. Rather than proceed in this fashion, we shall guess the range and then verify it. We recall that

$$\lambda = 2/\prod_{j=-\frac{s}{2}}^{s/2} |2\sin\tfrac{1}{2}k_j|,$$

so that, as in our solution with modified boundary conditions, the maximum λ will be obtained by fitting as many k_j's with $|\sin \frac{1}{2} k_j| \leq \frac{1}{2}$ or $|k_j| \leq \frac{\pi}{3}$ as possible. Thus we expect that for the maximum eigenvalue, where $\frac{S}{N} = \frac{1}{2}$, we will have $-\frac{\pi}{3} \leq k \leq \frac{\pi}{3}$. Our integral equation now becomes

$$\pi\, \rho(k) = 1 + \int_{-\pi/3}^{\pi/3} \rho(q) \operatorname{Re} \{\frac{1}{1+e^{ik}(e^{iq}-1)} - \frac{1}{2}\} dq.$$

In order to solve this equation, we attempt to write the integral as a convolution. This may be done by making the change of variable

$$e^{ik} = e^{\frac{\pi i}{3}} \frac{e^{\alpha} + e^{-\frac{\pi i}{3}}}{e^{\alpha} + e^{\frac{\pi i}{3}}}, \quad e^{iq} = e^{\frac{\pi i}{3}} \frac{e^{\beta} + e^{-\frac{\pi i}{3}}}{e^{\beta} + e^{\frac{\pi i}{3}}},$$

the limits now becoming $-\infty$ to $+\infty$. We then find

$$\pi\rho(k) = 1 + \frac{1}{2} \int_{-\infty}^{\infty} \rho(q) \frac{dq}{d\beta} \frac{\cosh \alpha + \frac{1}{2}}{\cosh (\alpha-\beta) + \frac{1}{2}} d\beta.$$

But $dq/d\beta = \dfrac{\frac{1}{2}\sqrt{3}}{\cosh \beta + \frac{1}{2}}$, and so we conclude that if

$$R(\beta) \equiv \rho(q) \frac{dq}{d\beta},$$

then

$$\frac{4\pi}{\sqrt{3}} R(\alpha) = \frac{2}{\cosh \alpha + \frac{1}{2}} + \int_{-\infty}^{\infty} \frac{R(\beta) d\beta}{\cosh (\alpha-\beta) + \frac{1}{2}}.$$

The solution now follows at once by Fourier transformation. We define

$$S(\gamma) = \int_{-\infty}^{\infty} e^{i\gamma\alpha} R(\alpha) d\alpha$$

$$Y(\gamma) = \int_{-\infty}^{\infty} e^{i\gamma\alpha} \frac{d\alpha}{\cosh \alpha + \frac{1}{2}} = \frac{4\pi}{\sqrt{3}} \frac{\sinh (\frac{\pi}{3} \gamma)}{\sinh (\pi\gamma)},$$

so that

$$\frac{4\pi}{\sqrt{3}} S(\gamma) = 2\,Y(\gamma) + Y(\gamma)S(\gamma)$$

or

$$S(\gamma) = \frac{2 \sinh\left(\frac{\pi\gamma}{3}\right)}{\sinh(\pi\gamma) - \sinh\left(\frac{\pi\gamma}{3}\right)} = \frac{1}{\cosh\left(\frac{2\pi}{3}\gamma\right)} .$$

An immediate consequence is that

$$\int_{-\pi/3}^{\pi/3} \rho(k)\,dk = \int_{-\infty}^{\infty} R(\alpha)\,d\alpha = S(0) = 1,$$

which verifies our choice of the range of k. Reversing the Fourier transform we now obtain the complete solution

$$\rho(\alpha) = \frac{1}{2\pi}\int_{-\infty}^{\infty} e^{-i\alpha\gamma}\, S(\gamma)\,d\gamma = \frac{3}{4\pi \cosh \frac{3\alpha}{4}} .$$

There remains the task of showing that the corresponding eigenvector has strictly positive components, in order that the eigenvalue of the transfer matrix will be maximum. For this purpose we note (see Yang and Yang, Phys. Rev. 150, 321-1966) that the anisotropic Heisenberg model satisfies the conditions of the Perron-Frobenius theorem when the coefficient Δ of $S_i^z S_{i+1}^z$ is $\frac{1}{2}$. By continuity from $\Delta = 0$, the Bethe ansatz with our choice of the k_j can be shown to yield the maximum eigenvalue at $\Delta = \frac{1}{2}$, and it then follows that the eigenvector has positive components.

Finally, let us compute the characteristic quantity W for the square ice model. We have

$$\log W = \frac{1}{N}\log\lambda = -\frac{1}{N}\sum_{j=-s/2}^{s/2} \log\left|2 \sin \frac{1}{2} k_j\right|$$

$$= -\frac{1}{2}\int_{-\pi/3}^{\pi/3} \rho(k)\,\log\left|2\sin\frac{1}{2}k\right|\,dk$$

$$= \frac{1}{4}\int_{-\infty}^{\infty} R(\alpha)\,\log\left(\frac{1 - e^{-3\alpha}}{(1 - e^{-\alpha})^3}\right)d\alpha,$$

or going over to Fourier transform

$$\log W = \frac{1}{8}\int_{-\infty}^{\infty} S(\gamma)\ (3\ \coth \pi\gamma - \coth \frac{\pi}{3}\gamma)\frac{d\gamma}{\gamma}.$$

Inserting $S(\gamma)$ and making the transformation $e^{\frac{2\pi\gamma}{3}} = x$ we get

$$\log W = \frac{1}{2}\int_{0}^{\infty} \frac{1}{\log x} \cdot \frac{x^2-1}{x^2+x+1} \cdot \frac{dx}{1+x^2}.$$

This is readily evaluated (see e.g. Gradshteyn and Ryzhik 4.267 No. 18, 19) yielding $\frac{3}{2}\log\frac{4}{3}$. We conclude that

$$W = (\tfrac{4}{3})^{3/2}.$$

4§. Other Hydrogen Bonded Models - Dimer Solution.

In the ice model just considered, each configuration was given a weight of either 0 or 1. In statistical mechanics, this generally represents a limiting situation of either 0 or infinite temperature. At finite temperature T, the weight of a configuration Γ is determined by the energy of the configuration

$$\omega(\Gamma) = e^{-\beta E(\Gamma)}, \quad \beta = \frac{1}{kT},$$

and after constructing the configuration sum (partition function)

$$Z = \sum_{\Gamma} \omega(\Gamma),$$

the standard quantity of interest is the free energy per site, f:

$$\beta f \equiv \lim_{n\to\infty} -\frac{1}{n}\log Z(n),$$

where n is the number of sites. In terms of the "number of combinations per site" $\log W = \lim_{n\to\infty}\frac{1}{n}\log Z(n)$ introduced previously, we have

$$W = e^{-\beta f}.$$

If $E(\Gamma)$ has only two values, 0 for an allowed configuration, ∞ for one which is disallowed, then Z reduces to a counting problem and W has precisely the same significance as before. However, if intermediate energies occur, W depends explicitly upon the inverse temperature β, and the analytic nature of this dependence is the primary object of mathematical and physical interest.

A hydrogen bonded structure similar to that of ice occurs in a number of crystals and is responsible for the oddity of their behavior. Prominent among these are the ferroelectrics. Let us first recall that a ferromagnetic crystal consists essentially of elementary magnetic dipoles with discrete possible directions which are correlated both by an external magnetic field and by their own net internal magnetic field. Regions of similar orientation have lower energy, and when the temperature is decreased (β increased) past the "Curie point" T, even an infinitesimal applied external field will cause a finite difference in the fraction of the dipoles pointing in its direction, rather than the reverse; an analytic break occurs in the free energy and other associated quantities at this temperature as well. Now a ferroelectric crystal differs mainly in that it does not have little electric dipoles, but rather charges -- hydrogen ions in the case we consider -- which can move to form polarized regions of lower energy, so that at low enough temperature a finite polarization occurs, signaled by analytic breaks in everything.

In the Slater model of KDP (potassium dihydrogen phosphate, KH_2PO_4),

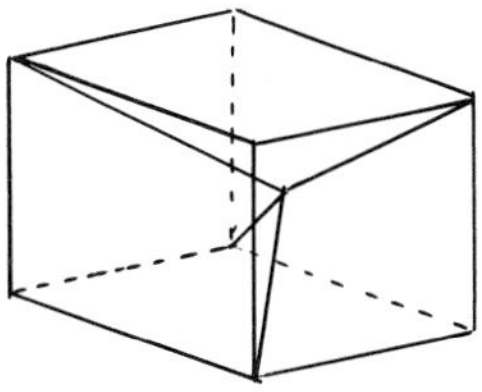

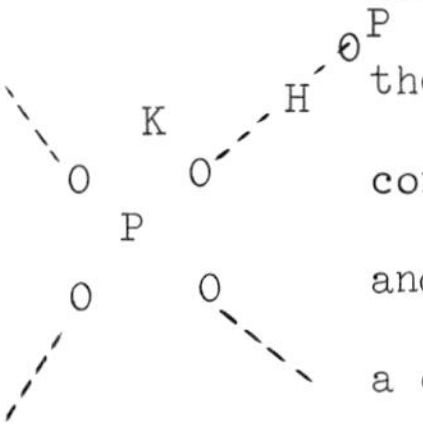

a lattice of P's is connected as in ice, with 4 O's next to each P on their respective arms. A hydrogen bond connects the two O's along each arm, and K is off center, establishing a characteristic crystal direction. The positions of the hydrogen ions are restricted by the "ice condition" that two H's must be close to each P, two H's far from it.

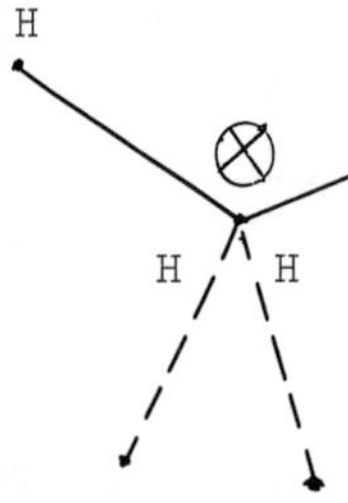

It is easy to see that each of the six "ice" possibilities results in a center of charge ⊗ which is along one of the six cubic directions. The PK axis is of course a special one, and in the Slater model we associate an energy of zero to either motion along the PK axis, of ε to the four remaining motions. A somewhat simpler model to solve (modified KDP of F. Y. Wu Phys. Rev. <u>168</u>, 539-1968) prohibits one of the two PK directions -- takes its energy as infinity. Projecting onto the plane as we did for ice, we have

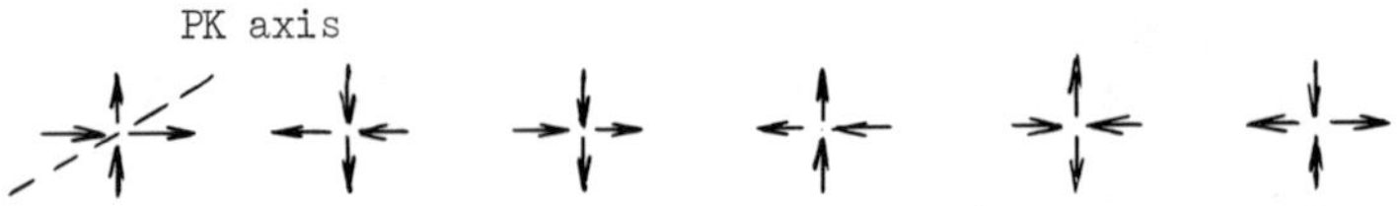

Slater KDP energy	0	0	ε	ε	ε	ε
Wu KDP energy	∞	0	ε	ε	ε	ε
Wu KDP weight	0	1	$u \equiv e^{-\beta\varepsilon}$	u	u	u

It is to be noted that in the Wu model, only one direction of polarization along the PK axis is permitted, so that the "Curie point" is that temperature at which the relative occupation

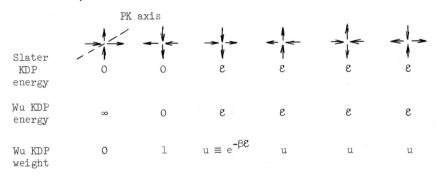

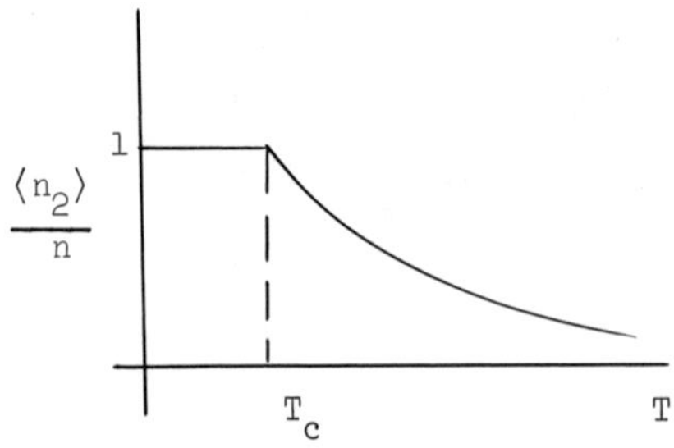

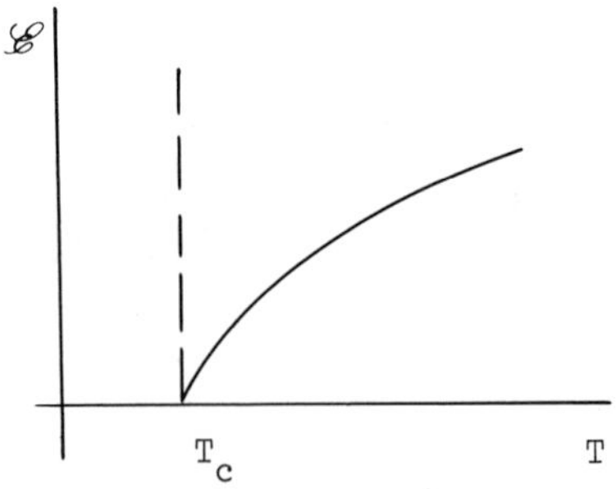

$\langle n_2\rangle/n$ of this configuration becomes unity: the dipole orientation snaps into position ($\langle n_2\rangle/n$ of course $\to 1$ as $T \to 0$). We will defer examination of the

degree of polarization until later (see the Ising model discussion) and at this stage recognize the Curie point by the nonanalyticity of the free energy, or to sharpen the effect, of the mean energy

$$\mathscr{E} \equiv \frac{1}{n} \sum E(\Gamma)\omega(\Gamma)/Z = \frac{\partial}{\partial\beta} \beta f.$$

Let us then compute the partition function

$$Z = \sum_{\Gamma} \exp\,(-\beta \sum_{j} E_j(\Gamma)) = \sum_{\Gamma} (\prod_{j} \omega_j(\Gamma))$$

for the N by N Wu model, where $E_j(\Gamma)$ denotes the energy of the j-th site in configuration Γ and $\omega_j(\Gamma)$ the corresponding weight. We observe that for the five allowed patterns, the weight at a given site is in fact the product of the weights of its associated arrows, namely,

$$\omega(\downarrow) = 1 \qquad \omega(\uparrow) = u^{1/2}$$

$$\omega(\leftarrow) = 1 \qquad \omega(\rightarrow) = u^{1/2}.$$

Thus we may write

$$Z = \sum_{\Gamma} \prod_{\alpha} \omega_\alpha(\Gamma),$$

where ω_α is the weight of the arrow at location α. The only problem then is to restrict Γ to the allowed configurations. Perhaps the simplest way of doing this is to associate a dimer packing with each allowed configuration. As the first step, we replace the up and right arrows, the only ones with weight other than unity, by dimers (each contributes to two sites), so that the allowed patterns become

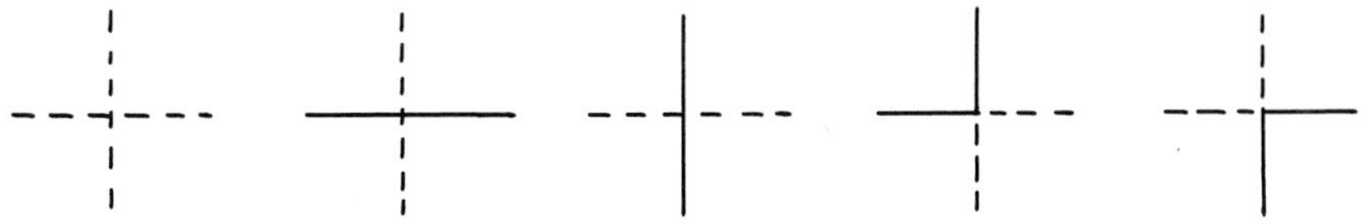

with a weight of u for each dimer present. However, two dimers overlap in 4 of the 5 cases. To avoid this, we decorate each vertex by separating it into two vertices, thereby converting the five patterns to

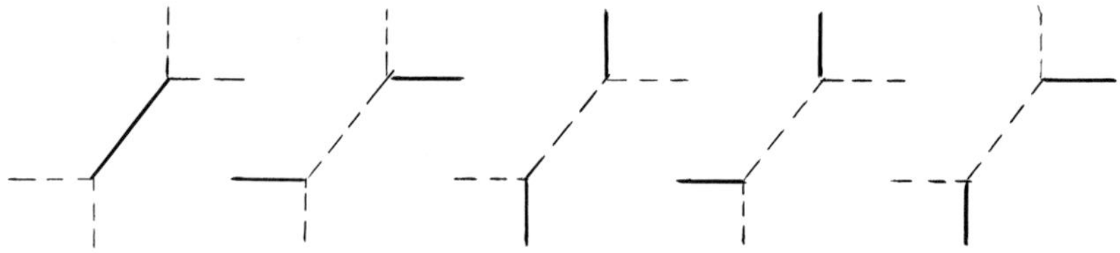

Now all patterns have nonoverlapping dimers, the weight of each horizontal or vertical arm is unchanged, the weight of each pattern is unchanged if a diagonal dimer is given weight 1, and the 5 patterns constitute all possible dimer packings on a site.

Thus our problem becomes that of a weighted dimer covering of a hexagonal or honeycomb lattice. To solve this as rapidly as possible, we distort each hexagon into a 6-vertex rectangle, and now the lattice becomes a brick design:

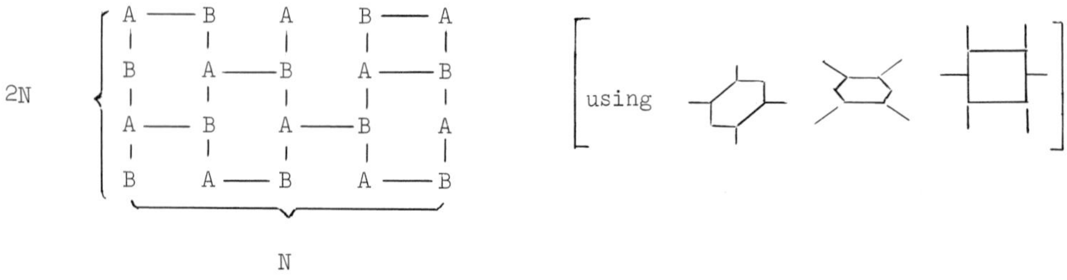

with weight 1 for horizontal dimers, u for vertical dimers and of course 0 for prohibited bonds. By taking the product of two configurations, one with each dimer given an $A \to B$ direction, the other with $B \to A$, we have converted the covering problem to that of oriented closed loops on a brick design lattice. Hence, assuming periodic boundary conditions, we have

$$Z^2 = \text{Per } K$$

where

$$K = \begin{pmatrix} C & uI & & & & uI \\ uI & D & uI & & & \\ & uI & C & \ddots & & \\ & & \ddots & D & \ddots & \\ & & & \ddots & \ddots & \ddots \\ uI & & & & \ddots & \ddots \end{pmatrix}_{2N \times 2N}$$

and

$$C = \begin{pmatrix} 0 & 1 & & & & \\ 1 & 0 & & & & \\ & & 0 & 1 & & \\ & & 1 & 0 & & \\ & & & & \ddots & \\ & & & & 0 & 1 \\ & & & & 1 & 0 \end{pmatrix}_{N\times N} \qquad D = \begin{pmatrix} 0 & & & & & 1 \\ & 0 & 1 & & & \\ & 1 & 0 & & & \\ & & & 0 & 1 & \\ & & & 1 & 0 & \\ & & & & & \ddots \\ 1 & & & & & 0 \end{pmatrix}_{N\times N}$$

are the connection matrices from a row to itself. The parity of a loop configuration is given, just as in the case of a dimer covering on a full square lattice, by

$$(-1)^P = i^{\Sigma v(R)}$$

where $\Sigma v(R)$ is the total number of vertical links which are occupied. Therefore

$$\text{Per } K = \text{Det } \overline{K}$$

where

$$\overline{K} = \begin{pmatrix} C & iuI & & & iuI \\ iuI & D & iuI & & \\ & iuI & C & \ddots & \\ & & \ddots & \ddots & \ddots \\ iuI & & & \ddots & \ddots \end{pmatrix}_{2N \times 2N}$$

To evaluate $\text{Det } \overline{K}$, we first reduce it by multiplying every even column on the right by C; then adding -iu times the sum of the two adjacent off columns. We obtain

$$\text{Per } K = \text{Det } \bar{\bar{K}}$$

where

$$\bar{\bar{K}} = \begin{pmatrix} DC+2u^2I & u^2I & 0 & & u^2I \\ u^2I & DC+2u^2I & u^2I & \ddots & \\ & u^2I & DC+2u^2I & \ddots & \\ & \ddots & \ddots & \ddots & u^2I \\ u^2I & & & u^2I & DC+2u^2I \end{pmatrix}_{N\times N}$$

and

$$DC = \begin{pmatrix} 0 & 0 & 0 & 0 & & \\ 0 & 0 & 0 & 1 & & \\ 1 & 0 & 0 & 0 & & \\ 0 & 0 & 0 & 0 & 0 & 1 \\ 0 & 0 & 1 & & & \\ & & & & & \end{pmatrix}_{N\times N} \qquad \begin{matrix} 2 \to 4 \\ 3 \to 1 \\ 4 \to 6 \\ 5 \to 3 \\ \vdots \end{matrix}$$

Let us find the eigenvalues of $\bar{\bar{K}}$. If v is an eigenvector then

$$\lambda v_{s,t} = (\bar{\bar{K}}v)_{s,t} = \begin{cases} v_{s,t+2} + 2u^2 v_{s,t} + u^2(v_{s+1,t}+v_{s-1,t}), & t \text{ odd;} \\ v_{s,t-2} + 2u^2 v_{s,t} + u^2(v_{s+1,t}+v_{s-1,t}), & t \text{ even;} \end{cases}$$

or, setting

$$t = \begin{cases} 2r+1 \\ 2r \end{cases} \quad \text{and} \quad \begin{matrix} v^+_{s,r} \equiv v_{s,2r+1} \\ v^-_{s,r} \equiv v_{s,2r} \end{matrix}$$

then

$$\lambda v^{\pm}_{s,r} = v^{\pm}_{s,r\pm1} + 2u^2 v^{\pm}_{s,r} + u^2(v^{\pm}_{s+1,r}+v^{\pm}_{s-1,r}).$$

Since $\bar{\bar{K}}$ is now periodic and translation invariant the $(k\ell)$-th eigenvector must be given by

$$v^{\pm}_{s,r}(k\ell) = e^{\frac{2\pi i}{N} sk} \cdot e^{\frac{2\pi i}{N/2} r\ell}$$

where $k = 1,\ldots,N$, $\ell = 1,\ldots,N/2$. We conclude that

$$\lambda^{\pm}_{k\ell} = e^{\pm i\theta} + 2u^2 + 2u^2 \cos \phi$$

where

$$\phi = \frac{2\pi}{N} K, \quad \theta = \frac{4\pi}{N}\ell.$$

In the limit $N \to \infty$, since $d\phi = 2\pi/N$, $d\theta = 4\pi/N$, we have

$$Z = (\mathrm{Det}\ \bar{\bar{K}})^{1/2} = \exp \frac{1}{2} \sum_{k,\ell} (\log \lambda^{+}_{k\ell} + \log \lambda^{-}_{k\ell})$$

$$\to \exp \frac{1}{2} \frac{N^2}{8\pi^2} \int_0^{2\pi} \int_0^{2\pi} 2 \log(e^{i\theta} + 2u^2 + 2u^2 \cos \phi) d\theta\, d\phi.$$

Rather than evaluate Z, we inspect the mean energy:

$$\mathscr{E} = -\frac{1}{N^2} \frac{\partial}{\partial \beta} \log Z$$

$$= \frac{\varepsilon}{8\pi^2} \int_0^{2\pi} \int_0^{2\pi} \frac{(2u^2+2u^2\cos \phi)}{e^{i\theta}+2u^2+2u^2\cos \phi}\, d\phi\, d\theta$$

Making the transformation $e^{-i\theta} = z$,

$$\mathscr{E} = -\frac{i\varepsilon}{8\pi^2} \int_0^{2\pi} \oint \frac{dz}{z + \frac{1}{2u^2+2u^2\cos \phi}}\, d\phi,$$

where the contour is the unit circle. Hence

$$\mathscr{E} = \frac{\varepsilon}{2\pi} \int_0^{2\pi} \begin{Bmatrix} 0 & \text{if} & 2u^2(1+\cos \phi) < 1 \\ 1 & \text{if} & 2u^2(1+\cos \phi) > 1 \end{Bmatrix} d\phi$$

$$= \begin{cases} 0 & \text{if } u = e^{-\varepsilon/kT} < \frac{1}{2}, \\ \frac{2\varepsilon}{\pi} \arccos \frac{1}{2} e^{\varepsilon/kT} & \text{if } u = e^{-\varepsilon/kT} > \frac{1}{2}, \end{cases}$$

the graph of which has been drawn above. The critical temperature in this case is given by

$$T_c = \frac{\varepsilon}{k \log 2} .$$

E. The Ising Model

1§. Introduction.

In the Ising model we associate with each site j on the lattice a variable s_j which can take on two values

$$s_j = \pm 1 .$$

In the most common realization, s_j represents the orientation of an atomic magnetic dipole with respect to some standard direction, +1 for the same direction, -1 for the opposite. In this case, because of the electronic origin of the dipole, s_j is referred to as spin. In another realization, the lattice is a binary alloy and $s_j = +1$ means that element A is at site j, $s_j = -1$ means that B is at j. Still another example is that of a lattice gas in which $s_j = +1$ if there is a gas molecule at site j, $s_j = -1$ if there is not. We will concentrate on the magnetic dipole interpretation, mainly because the external magnetic field which enters this system in a very natural way serves as a convenient parameter for analyzing the properties of the system.

The distribution or weight of configurations in a statistical ensemble depends upon the energy of each configuration. A general form for this energy is

$$E\{s_j\} = \frac{1}{2} \sum_{i \neq j} V_{ij} (1 - s_i s_j) - \mu B \sum_j s_j$$

where B is the value of the applied magnetic field and μ is the dipole moment at each site. In the ferromagnetic Ising model, the energy is lowered when two spins are parallel; thus $V_{ij} \geq 0$ and in fact our attention will be devoted almost solely to nearest neighbor interactions:

$$V_{ij} = \begin{cases} V & \text{if } i \text{ and } j \text{ are nearest neighbors} \\ 0 & \text{otherwise.} \end{cases}$$

At any rate, the weight of a configuration $\{s_j\}$ at temperature $T = 1/k\beta$ is again given by

$$\omega\{s_j\} = e^{-\beta E\{s_j\}}$$

and the weighted total number of configurations

$$Z = \sum_{\{s_j = \pm 1\}} \omega\{s_j\}.$$

The corresponding free energy

$$F = -\frac{1}{\beta} \log Z$$

then determines everything of consequence, from the mean energy

$$E = \langle E\{s_j\}\rangle = \sum_{\{s_j\}} E\{s_j\}\omega\{s_j\} \frac{1}{Z} = \frac{\partial}{\partial\beta} \beta F$$

and specific heat

$$C = \frac{\partial E}{\partial T} = -k\beta^2 \frac{\partial^2}{\partial\beta^2} [\beta F]$$

to the mean magnetic moment

$$\mathcal{M} = \langle \mu \sum_j s_j \rangle = \sum_{\{s_j\}} \frac{1}{Z}(\sum_j \mu\ s_j)\omega\{s_j\}$$

$$= -\frac{1}{\beta} \frac{\frac{\partial}{\partial B} Z}{Z} = \frac{\partial F}{\partial B}$$

and susceptibility

$$\mathcal{X} = \frac{\partial \mathcal{M}}{\partial B} = \frac{\partial^2 F}{\partial B^2}$$

The question before us is that of the dependence of the thermodynamics, as determined by F, on the parameters β and B. The quantities, E, C, $\mathcal{M}$, $\mathcal{X}$ which we have defined not only represent the possibility of direct measurement, but also as derivatives accentuate any singular behavior which F may have.

To anticipate, at very high temperature (very low β) all configurations have equal weight so that the two possible values of the spins will be equally distributed. At lower temperature, the lower energy associated with $s_j = \operatorname{sgn}(B)$ will give higher weight to configurations which have a preponderance of this sign of spin. At zero temperature $(\beta = \infty)$ only the single configuration with all spins of $\operatorname{sgn}(B)$ will occur. Correspondingly, the normalized magnetic moment or absolute magnetization

$$I = |\mathcal{M}\frac{1}{n\mu}| = |\frac{1}{n}\langle \sum_{j=1}^{n} s_j\rangle|$$

rises from 0 to 1 during this process.

If we take the limit as $B \to 0$, then of course we will have $I = 0$ at high temperature and $I = 1$ at $T = 0^{\circ}K$. We might expect in fact that I remains 0 and then changes discontinuously at $T = 0$. However for the ferromagnetic Ising model, there is a temperature T_c -- the curie temperature -- below which even an infinitesimal B will produce a finite non-zero value of I. i.e. will tend to line up or magnetize the atomic dipoles. This temperature also corresponds to a thermodynamic phase transition, manifesting itself as well in binary alloys. It may be mentioned that the nature of the permanent magnetization of a ferromagnet may be more complicated due to the possibility of a metastable collection of separately magnetized domains, but we will not be concerned with this process.

2§. Estimates of the Curie Temperature.

We want to find out under what circumstances the absolute magnetization per site $I = |\frac{1}{n}\langle \sum_{j=1}^{n} s_j\rangle|$ can remain nonzero as the external field $B \to 0$, i.e. when do we have a permanent magnet? Here we keep in mind an infinite system: the limit $n \to \infty$ is to be taken before questions are asked. Now without B, the energy $E\{s_j\} = \frac{1}{2}\sum_{i\neq j} V_{ij}(1 - s_i s_j)$ is unchanged under the transformation $s_j \to -s_j$ for all j, thus the weight $\omega\{s_j\}$ is unchanged, and we must have $\langle \sum_{j=1}^{n} s_j\rangle = \langle \sum_{j=1}^{n} -s_j\rangle$ or necessarily $\frac{1}{n}\langle \sum_{j=1}^{n} s_j\rangle = 0$. If the spins are independent (e.g. if $\beta V = 0$), this would occur in detail because the width of the distribution

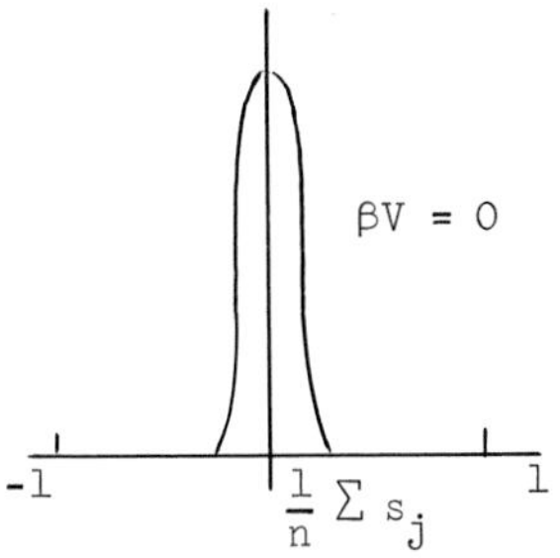

of the average spin would go as $n^{-1/2}$. On the other hand, with interaction, $\beta V > 0$, there is a tendency for neighboring spins to be parallel and hence a collective tendency for graphs of spins to be parallel. Thus we may expect a distribution like that shown: What happens below the Curie $T < T_c$ -- if one exists -- is that the tendency to line up is so strong that in essentially all configurations, some fraction of the spins are lined up with remaining spins nearly random; thus we have:

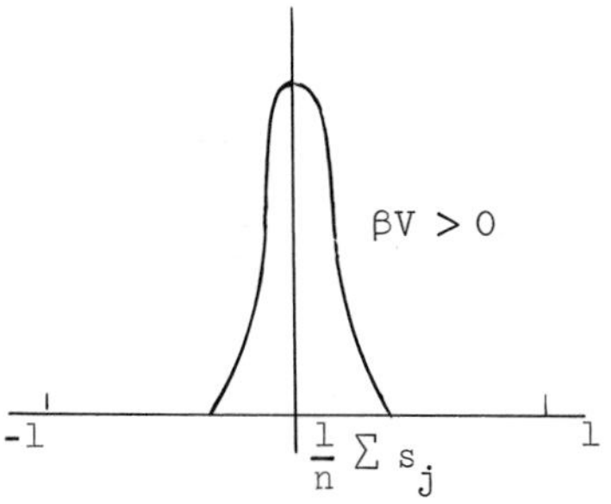

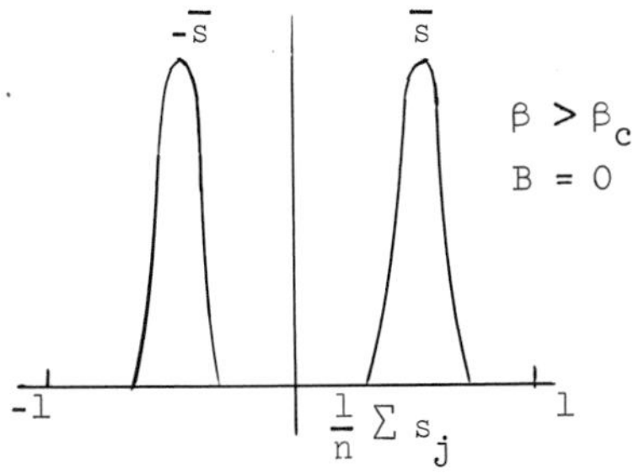

Under these circumstances, the effect of an external field will be dramatic: the weight factor introduced is $e^{\beta\mu B\,\Sigma s_j}$ which becomes $e^{\beta\mu n|\bar{s}B|}$ for the field-parallel case, $e^{-\beta\mu n|\bar{s}B|}$ for the net spin against the field.

If $n \to \infty$ first the weight difference is infinite for <u>any</u> value of $B \neq 0$, and so

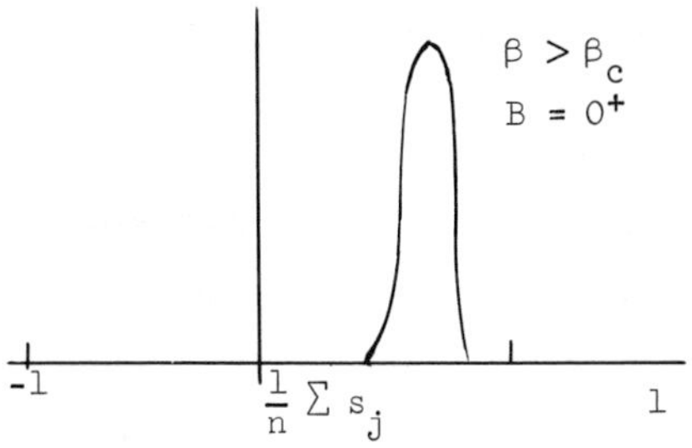

We see then that permanent magnetization can be assessed without actually applying a field. We simple have to show that $\frac{1}{n}\langle|\Sigma s_j|\rangle$ (or alternatively $\frac{1}{n^2}\langle(\Sigma s_j)^2\rangle$ has a finite nonzero value, rather than considering

$\frac{1}{n}\langle\sum s_j\rangle$ which necessarily vanishes. Let us examine the two-dimensional Ising model on a periodically bounded square lattice with nearest neighbor potential V. Then we have a weight

$$v = e^{-2\beta V}$$

for opposite spin neighbors, a weight of unity for parallel spin neighbors. We now proceed with an analysis

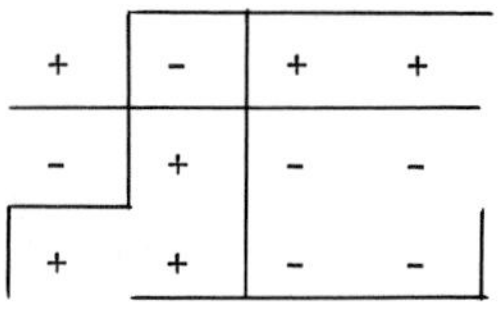

due to R. Peierls (Proc. Camb. Phil. Soc. 32, 477, 1936) and R. B. Griffiths (Phys. Rev. 136, A437, 1964). Suppose we place each spin in a cell and draw the cell boundary when it separates opposite spins. This decomposes the lattice into identical spin regions surrounded by closed curves. If the sense of each curve is chosen such that the spin of least total occupation is on the inner periphery, a decomposition into non-intersecting closed curves is obtained. Clearly, if L is the total perimeter of all curves and $g(L)$ the number of configurations of total perimeter L, the configuration sum is given by

$$Z = \sum_L g(L)v^L.$$

It is often useful to make a more detailed partition of Z: if $\{C_{\alpha,p}\}$ is a listing of all possible simple closed curves of perimenter p on the lattice, and

$$X_{\alpha,p}\{s_j\} = \begin{cases} 1 & \text{if it is one of the curves in the graph of } \{s_j\}, \\ 0 & \text{otherwise,} \end{cases}$$

then

$$Z = \sum_{\{s_j\}} v^{\sum_{\alpha,p} p\, X_{\alpha,p}\{s_j\}}.$$

Now the coefficient of magnetization may be written as

$$\mathcal{T} = \frac{1}{n} \langle | \Sigma s_j | \rangle$$
$$= \frac{1}{n} \langle | n_+ - n_- | \rangle$$
$$= 1 - \frac{2}{n} \langle \min (n_+, n_-) \rangle ,$$

where n_+ is the number of + spins, n_- of - spins. In order to find a lower bound for $\mathcal{T}$, we first observe that $\min (n_+, n_-)$ is certainly bounded from above by the total area $\mathcal{R}$ belonging to (i.e. that are which is $\leq n/2$) the curves $C_{\alpha,p}$ which belong to $\{s_j\}$. From the above discussion we have

$$\langle \mathcal{R} \rangle = \sum_{\{s_j\}} \sum_{\alpha,p} X_{\alpha,p}\{s_j\} \mathcal{R}_{\alpha,p} v^{\sum qX_{\beta,q}\{s_j\}}_{\beta,q} \frac{1}{Z} .$$

But $Xv^{pX} = Xv^p$ for $X = 0,1$ so that

$$\langle \mathcal{R} \rangle = \sum_{\{s_j\}} \sum_{\alpha,p} [X_{\alpha,p}\{s_j\} \mathcal{R}_{\alpha,p} v^{\sum_{(\beta,q)\neq(\alpha,p)} qX_{\beta,q}\{s_j\}} v^p] \frac{1}{Z} .$$

To estimate $\mathcal{R}_{\alpha,p}$, we see that for a curve which can be contracted to a point on the periodic square, $\mathcal{R}_{\alpha,p}$ is maximum at fixed p when the curve is a square so that $\mathcal{R}_{\alpha,p} \leq (p/4)^2$. On the other hand, if $C_{\alpha,p}$ cannot be contracted to a point, then $p \geq \sqrt{n}$ while $\mathcal{R}_{\alpha,p} \leq n/2$ so that $\mathcal{R}_{\alpha,p} \leq p^2/2$ -- which also includes the first case. Thus

$$\langle \mathcal{R} \rangle \leq \sum_{\alpha,p} v^p \frac{p^2}{2} \sum_{\{s_j\}} X_{\alpha,p}\{s_j\} v^{\sum_{(\beta,q)\neq(\alpha,p)} qX_{\beta,q}\{s_j\}} \frac{1}{Z} .$$

Since the excision of $C_{\alpha,p}$ from a configuration of curves containing it creates another valid configuration of curves (corresponding to reversing the sign of every spin interior to $C_{\alpha,p}$) we have

$$\sum_{\{s_j\}} X_{\alpha,p}\{s_j\} v^{\sum_{(\beta,q)\neq(\alpha,p)} qX_{\beta,q}\{s_j\}}$$
$$\leq \sum_{\{s_j\}} X_{\alpha,p}\{s_j\} \sum_{\{s_j\}} v^{\sum_{(\beta,q)\neq(\alpha,p)} qX_{\beta,q}\{s_j\}}$$

$$\leq \sum_{\{s_j\}} X_{\alpha,p}\{s_j\} \sum_{\{s_j\}} v^{\sum_{\beta,q} qX_{\beta,q}\{s_j\}} = Z \sum_{\{s_j\}} X_{\alpha,p}\{s_j\}$$

or

$$\langle \mathscr{R} \rangle \leq \sum_{\alpha,p} \frac{p^2}{2} v^p \sum_{\{s_j\}} X_{\alpha,p}\{s_j\}.$$

Finally, the number of curves in $\{s_j\}$ is bounded by the number of curves of length p which are possible on the lattice. Each such curve must start at some point, have a choice of 4 directions the first time, 3 each remaining time except for one the last time, and suffer a $2p$-fold repetition due to redundancy of the starting point and direction of the curve. Hence $\sum_{\alpha} X_{\alpha,p}\{s_j\} \leq \frac{4n\, 3^{p-2}}{2p}$, and we have

$$\langle \mathscr{R} \rangle \leq \sum_p np\, 3^{p-2} v^p.$$

The length must be even and ≥ 4. Summing over p we conclude that

$$\mathscr{T} \geq \frac{1}{(1-9v^2)^2} [1 - 18v^2 + 9v^4 + 324v^6],$$

so that the spontaneous magnetization $\mathscr{T} > 0$ necessarily occurs for

$$v \leq v_c = 0.3 \ldots .$$

If we attempt to generalize the proceding to a d dimensional lattice, then if p is the surface of a region of identical spins, it is easy to see that

$$\mathscr{R}_{\alpha,p} \leq \frac{1}{2} p^{\frac{d}{d-1}} .$$

Similarly a very high upper bound

$$\sum_{\alpha} X_{\alpha,p}\{s_j\} \leq \frac{n(2d-1)^{p-2}}{2p} 2d$$

is available. We thus find that if $d \geq 2$, there is always a critical value of

$v = e^{2V/kT}$ and hence a critical temperature below which spontaneous magnetization will take place. On the other hand, if $d = 1$, the above estimates give no bound at all, principally because the length of an interval is not bounded by the number of end points, which is always 2. However, in one dimension, a direct evaluation of $\mathscr{T}$ is readily made. We can write

$$\mathscr{T} = \frac{1}{n} \langle | n_+ - n_- | \rangle$$

$$= \frac{1}{n} \sum_{p, n_+, n_- | n_+ + n_- = n} | n_+ - n_- | v^p \mathscr{N}(n_+, n_-, p)$$

$$\Big/ \sum_{p, n_+, n_- | n_+ + n_- = n} v^p \mathscr{N}(n_+, n_-, p),$$

where $\mathscr{N}(n_+, n_-, p)$ is the number of configurations $\{s_j\}$ such that

$$\frac{1}{2} \sum_j (1 + s_j) = n_+ \quad = \text{number of } + \text{ spins}$$

$$\frac{1}{2} \sum_j (1 - s_j) = n_- \quad = \text{number of } - \text{ spins}$$

$$\frac{1}{2} \sum_j (1 - s_j s_{j+1}) = p = \text{number of reversals.}$$

Let us compute $\mathscr{N}(n_+, n_-, p)$ on a lattice with fixed end points (not periodic) when p is odd. To do this, suppose that $s_1 = \pm 1$; we contact all runs of $+1$'s, giving a sequence of n_+ + 1's in $\frac{p+1}{2}$ "boxes". This can be done in $\binom{n_+ - 1}{\frac{p+1}{2} - 1}$ ways (see p. 34ff). Similarly, the grouped -1's give rise to $\binom{n_- - 1}{\frac{p-1}{2}}$ combinations. Since an allowed configuration is produced by alternating runs of $+1$'s and -1's, there are $\binom{n_+ - 1}{\frac{p-1}{2}}\binom{n_- - 1}{\frac{p-1}{2}}$ configurations starting with a $+1$; the same number is created by starting with -1, and so we have

$$\mathscr{N}(n_+, n_-, p) = 2 \binom{n_+ - 1}{\frac{p-1}{2}} \binom{n_- - 1}{\frac{p-1}{2}}.$$

This is strictly true for p odd, and must be modifed slightly for even p. However, in the asymptotic limit $n_+, n_-, p \to \infty$ we easily find for both cases

$$\mathcal{N}(n_+,n_-,p) = \frac{2^{n+2}}{\pi\,(n_+n_-)^{1/2}}\; e^{-\frac{n}{2}\left(1-\frac{4n_+n_-}{n^2}\right)}\; e^{-\frac{1}{2}\left(\left(\frac{n}{n_+n_-}\right)^{1/2} p-2\left(\frac{n_+n_-}{n}\right)^{1/2}\right)^2} .$$

It follows that

$$\sum_p v^p \mathcal{N}(n_+,n_-,p) = \frac{2^{n+2}}{(2n\pi)^{1/2}}\; v^{\frac{2n_+n_-}{n}}\; e^{-\frac{n}{2}\left(1-\frac{4n_+n_-}{n^2}\right)} ;$$

setting $n_+ = n/2 + t$, $n_- = n/2 - t$ we thus have

$$\mathcal{T} = \langle \tfrac{2}{n}|t| \rangle = \sum_t 2|t|\,(ev)^{-2t^2/n} \Big/ n \sum_t (ev)^{-2t^2/n}$$

$$= \left(\frac{2}{n}\right)^{1/2}\left(\frac{2}{\pi\,(1+\log v)}\right)^{1/2}.$$

We conclude that $\mathcal{T} \to 0$ as $n \to \infty$, so that no permanent magnetization exists for the one dimensional Ising model.

A much better estimate is obtainable by a Pauling-type approximation sequence due to Kikuchi (Phys. Rev. 81, 988, 1951). The general idea is to introduce intensive variables $\lambda_i\{s_j\}$ (i.e. a total quantity divided by the number of lattice sites) in terms of which the total energy can be written exactly, and then approximate the total weight corresponding to fixed values of $\lambda_i\{s_j\}$. Thus one writes

$$Z = \sum_{\{s_j\}} \omega\{s_j\} = \sum_{\lambda_1,\lambda_2,\ldots} e^{-\beta n \mathcal{E}(\lambda_1,\lambda_2,\ldots)} [w_n(\lambda_1,\lambda_2,\ldots)]^n ,$$

where $E\{s_j\} \equiv n\mathcal{E}(\lambda_1,\lambda_2,\ldots)$ and the weight w_n is independent of n for large n. As $n \to \infty$, only the maximum of the summand with respect to the λ_i will contribute, and in fact the free energy per site will be given by

$$f = \frac{F}{n} = -\frac{1}{n\beta}\log Z \to \mathcal{E} - \frac{1}{\beta}\log w$$

evaluated at the minimum of f, i.e.

$$\frac{df}{d\lambda_i} = 0, \quad f \text{ is minimum.}$$

The approximations which we will consider will relate only to

$$\log w = S/k \quad \text{(the entropy per site)}$$

since $\mathcal{E}(\lambda_1,\lambda_2,\ldots)$ is of course given explicitly. We observe that this is strictly a combinatorial problem, since our definition of w may be written as

$$w(\lambda_1',\lambda_2',\ldots) = \lim_{n\to\infty}\Big[\sum_{\lambda_i\{s_j\}=\lambda_i'\ i=1,2,\ldots} 1\Big]^{1/n}.$$

Consider the one dimensional Ising model. An adjacent pair of vertices can have four possible configurations: ++, +-, -+, --. As our intensive variables, we choose the relative frequencies of these configurations

$$\lambda_{\pm\pm} = \frac{1}{n}\sum_j \frac{1}{4}(1\pm s_j)(1\pm s_{j+1}).$$

Then indeed the energy $E = \frac{1}{2}\sum_j V(1-s_j s_{j+1}) = 2nV(\lambda_{+-} + \lambda_{-+})$, so that

$$\mathcal{E}(\lambda_{++},\lambda_{+-},\lambda_{-+},\lambda_{--}) = 2V(\lambda_{+-} + \lambda_{-+}).$$

The intensive variables we have chosen are of course not independent, for in addition to the obvious

$$\lambda_{++} + \lambda_{+-} + \lambda_{-+} + \lambda_{--} = 1,$$

we note that in terms of the relative frequencies of +'s and -'s

$$\lambda_{\pm} = \frac{1}{n}\sum_j \frac{1}{2}(1 \pm s_j),$$

we must have

$$\lambda_+ = \lambda_{++} + \lambda_{+-} = \lambda_{-+} + \lambda_{++}$$

$$\lambda_- = \lambda_{--} + \lambda_{-+} = \lambda_{+-} + \lambda_{--}$$

(because the number of +'s is the same as the number of pairs with a + on the left, etc.). We must now determine the number of configurations for which

$\mathcal{E}(\lambda_{++},\lambda_{+-},\lambda_{-+},\lambda_{--})$ have fixed values. Although we have already done so exactly for this particular lattice, it is instructive to apply Kikuchi's approach -- somewhat modified -- to this case as well.

If there were no restrictions on the sequence of pair configurations permitted aside from their relative frequencies, the number of lattice configurations would be given simply by the multinomial coefficient

$$\mathcal{N}^*(\lambda_{++},\lambda_{+-},\lambda_{-+},\lambda_{--}) = \frac{n!}{(n\lambda_{++})!(n\lambda_{+-})!(n\lambda_{-+})!(n\lambda_{--})!},$$

or applying the Stirling approximation and dropping multiplicative factors which $\to 1$ when the n-th root is taken as $n \to \infty$,

$$\mathcal{N}^*(\lambda_{++},\lambda_{+-},\lambda_{-+},\lambda_{--}) = \frac{1}{\lambda_{++}^{n\lambda_{++}}\lambda_{+-}^{n\lambda_{+-}}\lambda_{-+}^{n\lambda_{-+}}\lambda_{--}^{n\lambda_{--}}}.$$

But of course, not all sequences are permitted since the spin at the j-th vertex is both the right member of the $(j-1,j)$ pair and the left member of the $(j,j+1)$ pair, e.g. one cannot have a pair $(+,-)$ followed by $(+,+)$. Consider a $+$ vertex. Of the possible pairs of which it is the right member only $(+,+)$ and $(-,+)$ are allowed; thus the possible left pairs must be reduced by the factor $(\lambda_{++} + \lambda_{-+})/(\lambda_{++} + \lambda_{+-} + \lambda_{-+} + \lambda_{--}) = \lambda_+$, and similarly we correct for the extra right pairs by multiplying by λ_+. In the same way, a - vertex is corrected by multiplying by λ_-^2. For a configuration of $n\lambda_+$ +'s and $n\lambda_-$ -'s, we therefore require the correction

$$[\lambda_+^{n\lambda_+}\lambda_-^{n\lambda_-}]^2.$$

Since there are

$$\frac{n!}{(n\lambda_+)!(n\lambda_-)!} \approx \frac{1}{\lambda_+^{n\lambda_+}\lambda_-^{n\lambda_-}}$$

possible configurations which are to be corrected in this way, we conclude that

$$w = (\mathscr{N})^{1/n} = \frac{\prod_s \lambda_s^{\lambda_s}}{\prod_{s,s'} \lambda_{s,s'}^{\lambda_{s,s'}}},$$

and consequently

$$f = 2V(\lambda_{+-} + \lambda_{-+}) - \frac{1}{\beta}\sum_s \lambda_s \log \lambda_s + \frac{1}{\beta}\sum_{s,s'} \lambda_{s,s'} \log \lambda_{s,s'}.$$

In this special case, it can be shown that our approximate result for $\mathscr{N}^{1/n}$ in fact coincides asymptotically with the exact result $[\mathscr{N}(n_+, n_-, p)]^{1/n}$ previously found. There is of course in this case no Curie temperature at all, and the system remains in a "disordered" state -- i.e. the total spin is sharply distributed about zero. To find the free energy we must minimize f subject to all necessary restrictions. We will clearly have $\lambda_+ = \lambda_- = 1/2$, so that if $\lambda_{+-} = \lambda_{-+}$ defined as y then $\lambda_{++} = \lambda_{--} = 1/2 - y$. Hence

$$f = 4yV - \frac{1}{\beta}\log\frac{1}{2} + \frac{2}{\beta}[y \log y + (\tfrac{1}{2} - y)\log(\tfrac{1}{2} - y)],$$

with a minimum of

$$f_{min} = 2V - \frac{1}{\beta}\log(1 + e^{2\beta V}) \quad \text{at} \quad y_{min} = \frac{1}{2}\,\frac{1}{1 + e^{2\beta V}}.$$

As expected, f_{min} is analytic over the full range of β.

Let us now proceed to the two dimensional Ising model on a square lattice. To describe the system, we choose as parameters the relative frequencies $\lambda_+, \lambda_-, \lambda_{++}, \lambda_{+-}, \lambda_{-+}, \lambda_{--}, \lambda_{\substack{+\\+}}, \lambda_{\substack{+\\-}}, \lambda_{\substack{-\\+}}, \lambda_{\substack{-\\-}}$. As in the one dimensional case, the number of lattice configurations on the assumption of independent vertex pairs will be given by

$$\mathscr{N}^*(\lambda_{++}, \ldots, \lambda_{\substack{-\\-}}) = \frac{1}{\prod_{s,s'} \lambda_{s,s'}^{n\lambda_{s,s'}} \prod_{s,s'} \lambda_{\substack{s\\s'}}^{n\lambda_{\substack{s\\s'}}}}.$$

Then for each of $1/(\lambda_+^{n\lambda_+} \lambda_-^{n\lambda_-})$ configurations of +'s and -'s we have four conditions at each site, yielding the correction factor

$$[\lambda_+^{n\lambda_+} \lambda_-^{n\lambda_-}]^4$$

Since the system will be symmetric between horizontal and vertical directions, we can set $\lambda_{s \atop s'} = \lambda_{ss'}$, and therefore conclude that

$$w = [\mathcal{N}]^{1/n} = \frac{\prod_s \lambda_s^{3\lambda_s}}{\prod_{s,s'} \lambda_{s,s'}^{2\lambda_{s,s'}}}$$

which is known as the Bethe-Peierls approximation. The curve of the specific heat $c = -\beta^2 \frac{\partial^2}{\partial\beta^2} \beta f$ is given by the solid line for this approximation, and compared with the exact dashed line curve. The Curie temperature predicted in this way is given by $e^{-2\beta_c V} = \frac{1}{2}$, in comparison with the exact (see next section) $e^{-2\beta_c V} = \sqrt{2} - 1 = 0.414...$ and the Peierls-Griffiths lower bound $e^{-2\beta_c V} = 0.3...$. This approximation may be successively improved in the same fashion as was used for the Pauling treatment of the ice model. Thus, if consistency of pair configurations around a square rather than at a vertex is demanded we obtain the more accurate Kramers-Wannier approximation.

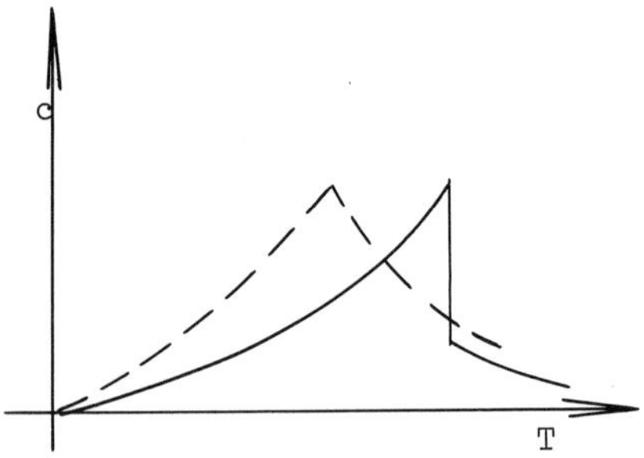

Let us conclude this section by pointing out the relationship between the existence of zero field permanent magnetization and long range spin correlation. As we have seen, such a state is specified by the fact that $\sum_j s_j$ is sharply peaked at $\pm nI$ where $0 < I < 1$. Thus $\langle \sum_{j=1}^{n} s_j \sum_{k=1}^{n} s_k \rangle = n^2 I^2$, or in terms of the spin-spin correlation coefficient

$$\langle s_j \, s_k \rangle = p_{j-k}$$

(for a very large lattice, only the vector difference of the spin sites can enter), $I^2 = \lim_{n \to \infty} \frac{1}{n^2} \sum_{j,k=1}^{n} p_{j-k}$. Hence

$$I^2 = \lim_{\ell \to \infty} p_\ell,$$

i.e. in an ordered state which permits permanent magnetization, spins infinitely far apart are not independent but instead possess a finite long range correlation.

We can translate the preceding to lattice gas language, in which $s_j = 2r_j - 1$. Thus the pair distribution

$$g_{j-k} = \frac{\langle r_j r_k \rangle}{\langle r_j \rangle \langle r_k \rangle}$$

becomes (using $\langle r_i \rangle = 1/2$, $\langle r_j r_k \rangle = \frac{1}{4} \langle (s_j+1)(s_k+1) \rangle$)

$$g_{j-k} = p_{j-k} + 1,$$

and in a state with long range order, we will not have $g_\ell \to 1$ as $\ell \to \infty$ -- which would be the case for asymptotically independent site occupations. In the case of the lattice gas, this is the typical effect of the simultaneous existence of two phases, i.e. suppose that the possible configurations divide up into two classes A and B with relative weights x and $1-x$, respectively. Class A has a uniform density $\langle r_j | A \rangle = \rho_1$ with no asymptotic correlations $\frac{\langle r_j r_k | A \rangle}{\langle r_j | A \rangle \langle r_k | A \rangle} \to 1$, while class B has uniform density ρ_2 and no asymptotic correlations. Clearly, $\rho_1 x + \rho_2(1-x) = \langle r_j \rangle = 1/2$ so that $x = \frac{\frac{1}{2} - \rho_2}{\rho_1 - \rho_2}$.

Then $\lim_{j-k \to \infty} g_{j-k} = \frac{\rho_1^2 x + \rho_2^2(1-x)}{[\rho_1 x + \rho_2(1-x)]^2} = 1 - (1-2\rho_1)(1-2\rho_2)$, which differs from unity unless $\rho_1 = \rho_2 = 1/2$.

3§. Combinatorial Solution of the Ising Model.

We have already considered the diagramatic expansion of the nearest neighbor Ising model partition function

$$Z = \sum_L g(L) v^L, \quad v = e^{-2\beta V}$$

where $g(L)$ is the number of L link decompositions of the lattice, a link marking the boundary between an adjacent pair of unlike spins. This is of course a low temperature expansion, converging rapidly for large β. One can write down a high temperature expansion in a very similar form. We observe that since $(s_i s_j)^2 = 1$, then $e^{\beta V s_i s_j} = \cosh \beta V + s_i s_j \sinh \beta V = (1 + u\, s_i s_j) \cosh \beta V$, where $u = \tanh \beta V$. It follows immediately that

$$\begin{aligned} Z &= \sum_{\{s_j\}} \prod_{n\cdot n\cdot p} e^{-\beta V(1 - s_i s_j)} \\ &= e^{-\frac{nc}{2}\beta V} \sum_{\{s_j\}} \prod_{n\cdot n\cdot p} e^{\beta V s_i s_j} \\ &= (\cosh \beta V)^{nc/2} e^{-\frac{nc}{2}\beta V} \sum_{\{s_j\}} \prod_{n\cdot n\cdot p} (1 + u s_i s_j), \end{aligned}$$

where $n\cdot n\cdot p$ refers to unordered nearest neighbor pairs (i,j), and c is the coordination number, the number of links to each vertex. Now expanding we have

$$\prod_{n\cdot n\cdot p} (1 + u\, s_i s_j) = 1 + u \sum_{n\cdot n\cdot p} s_i s_j + u^2 \sum_{n\cdot n\cdot p} s_i s_j \sum'_{n\cdot n\cdot p} s_k s_\ell + \ldots, \quad \sum'$$

indicating that $(k,\ell) \neq (i,j)$. Since $\sum_{s_i = \pm 1} = \sum_{s_i = \pm 1} s_i^3 = \ldots = 0$, each s_i in any term of the expansion must occur an even number of times up to c, or else the term will vanish on performing the sum over $\{s_j\}$. Each non-vanishing sum contributes a factor of 2, and so we conclude that

$$Z = 2^n e^{-\frac{nc}{2}\beta V} (\cosh \beta V)^{\frac{nc}{2}} \sum_r h(r) (\tanh \beta V)^r,$$

where $h(r)$ is the number of r-bond graphs on the lattice, with each vertex being of even order (having even number of bonds). As noted above, this is a high temperature expansion, converging rapidly for small β.

We shall shortly see how to sum each of the above series explicitly.

First, we will indicate how the exact Curie temperature may be obtained with very little effort merely by comparison of the two series. For this purpose, we shall assume that the Curie temperature -- that at which the various thermodynamic parameters are not analytic in β at zero applied field -- is unique. (This can be shown by using a theorem of Lee and Yang -- Phys. Rev. 87, 410, 1952 -- on the zeros of the grand partition function; see Newell and Montroll -- Rev. Mod. Phys. 25, 386, 1953).

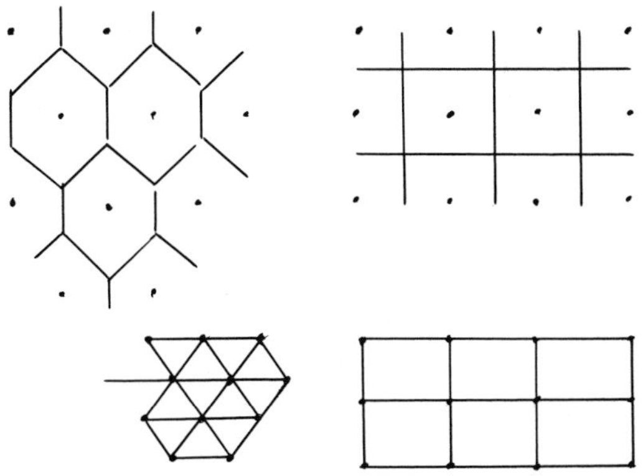

Suppose then that $Z(\beta) = \sum_L g(L)(e^{-2\beta V})^L$ is singular at $\beta = \beta_c$. Here $g(L)$ is the number of decompositions of the lattice into + regions and - regions, or perimeter L, which may also be regarded as the number of closed graphs of length L, with each vertex of even order, on the dual lattice shown on the first line of the accompanying diagram. We can transform partially to the high temperature form by defining

$$\tanh \beta^* V = e^{-2\beta V},$$

so that

$$Z(\beta) = \sum_L g(L)\ (\tanh \beta^* V)^L.$$

For the square lattice, either infinite or with periodic boundary conditions, the dual lattice is the lattice itself, whence $g(L) = h(L)$. Thus

$$Z(\beta^*) = 2^n\ e^{-2n\beta^* V}(\cosh \beta^* V)^{2n} Z(\beta).$$

Clearly $Z(\beta^*)$ is singular when $Z(\beta)$ is. We conclude that

$$\beta_c^* = \beta_c$$

or

$$e^{-2\beta_c V} = \tanh \beta_c V = \frac{1 - e^{-2\beta_c V}}{1 + e^{-2\beta_c V}} ,$$

which has the solution

$$e^{-2\beta_c V} = \sqrt{2} - 1 = 0.414\ldots .$$

For the triangular lattice, the dual is a honeycomb lattice and an additional transformation is then required to relate the two series -- see Newell and Montroll.

Let us now carry out in closed form the summation of the high temperature series for the zero field square lattice Ising model. We shall do this by a strictly combinatorial method based on converting the problem to a suitable dimer packing enumeration (M. Fisher, J. Math. Phys. 7, 1776, 1966). For this purpose, it will be helpful to reduce the configuration number $(c = 4$ for square lattice) to a minimum. Now in $Z = \sum_{\{s_j\}} \exp\{-\frac{1}{2}\beta \sum_{i \neq j} V_{ij}(1-s_i s_j)\}$, we observe that if any $V_{ij} \to \infty$, then only configurations for which $s_i = s_j$ will contribute to Z, i.e. the vertices i and j collapse to a single vertex. In the reverse way, by expanding a vertex as shown to $c-2$ vertices and $c-3$ extra bonds, but taking $V_{ij} \to \infty$ for each extra bond, the configuration number can always be reduced to 3.

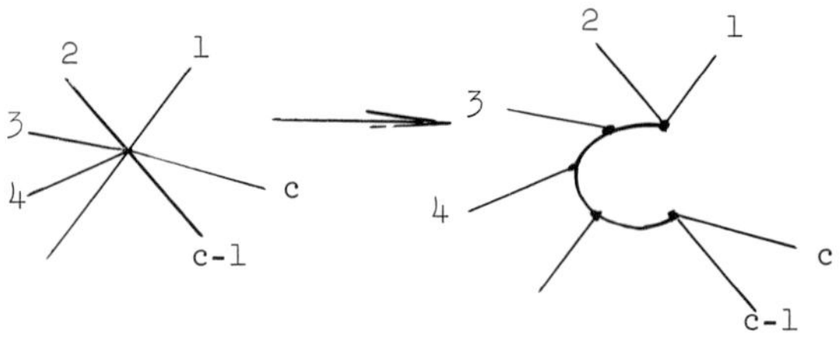

In the case of a square lattice, the "decoration" of each vertex thereby produced is simply

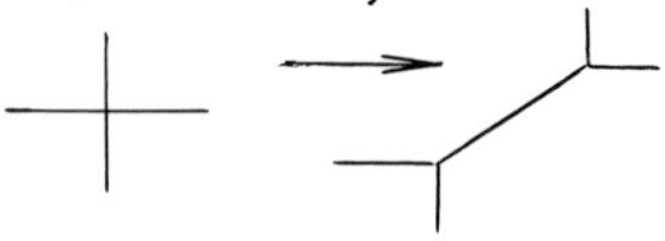

Since $e^{\beta V(s_i s_j - 1)} = e^{-\beta V} \cosh \beta V(1 + s_i s_j \tanh \beta V) \to 1 + s_i s_j$ as $V \to \infty$, the extra bonds each appear in the high temperature expansion with a weight of unity. Finally, the graphs on the decorated lattice with $c = 3$ must be put in one to one correspondence with dimer coverings. The simplest way to do this is to further decorate the lattice, as shown:

i.e. the absence of a link to a vertex corresponds to the presence of a dimer on the line leading to the expanded vertex. Thus the square Ising model becomes the dimer

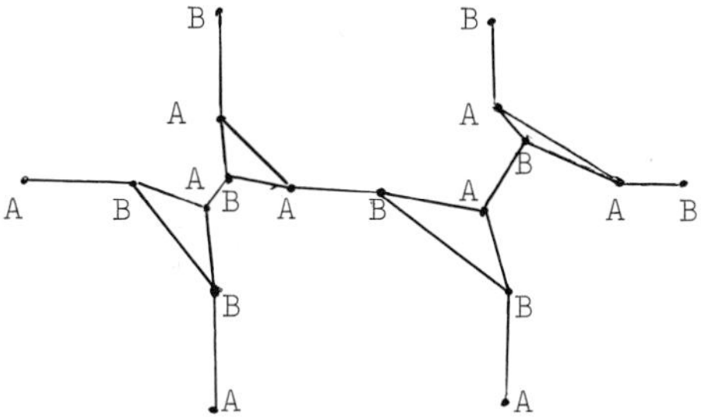

problem on a lattice in which each vertex decomposes into six vertices, and the various dimer locations are weighted in obvious fashion: $w = \coth \beta V$ for a horizontal or vertical link (becuase the presence of a bond corresponds to the absence of a dimer), $w = 1$ otherwise.

We now convert to a closed loop covering problem in essentially standard fashion: give all dimers of one packing on $A \to B$ orientation whenever this is possible, give all dimers of a second packing a $B \to A$ orientation, and fit the AA and BB arms by continuity. Thus, the square of the weighted number of configurations is expressed as a permanent. To convert this to a determinant, there is a general theorem of Kasteleyn: J. Math. Phys. 4, 287, 1963 (see proof at end of section).

Bond signs can be chosen on a planar graph such that every oriented cycle of even interior opposes an odd number of bonds. The appropriate signs for a square lattice, and for the decorated lattice under consideration are shown below. If K is the corresponding counting matrix, we will then have

$$Z = 2^n e^{-2n\beta V} (\sinh \beta V)^{2n} \det K$$

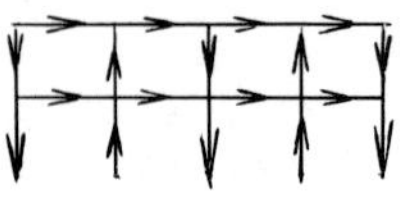

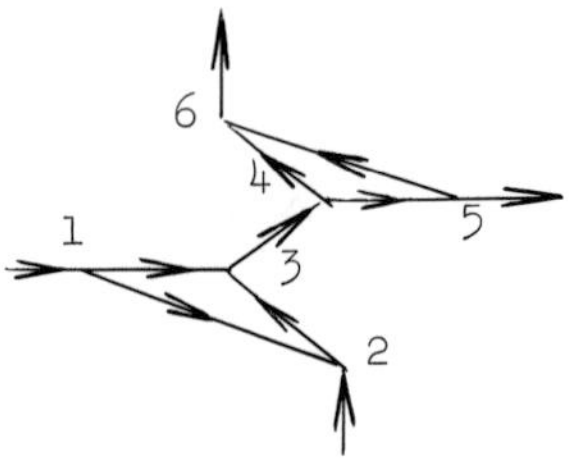

(because $(\cosh \beta V)^{2n} \sum_r h(r) \; (\tanh \beta V)^r = (\sinh \beta V)^{2n} \sum_r h(2n-r) \cdot (\coth \beta V)^r$, and $2n-r$ is the number of lines which are absent on the graph, and consequently the number of dimers).

The matrix K for a periodically bounded lattice falls into 6 by 6, as indicated:

$$K_{ij,[i'j']} = \begin{pmatrix} 0 & 1 & 1 & 0 & -w\omega_1^{-1} & 0 \\ -1 & 0 & 1 & 0 & 0 & -w\omega_2^{-1} \\ -1 & -1 & 0 & 1 & 0 & 0 \\ 0 & 0 & -1 & 0 & 1 & 1 \\ w\omega_1 & 0 & 0 & -1 & 0 & 1 \\ 0 & w\omega_2 & 0 & -1 & -1 & 0 \end{pmatrix}$$

where $\omega_1^{\pm 1}$ contributes only to the matrix element $(ij, i\pm 1\; j)$, $\omega_2^{\pm 1}$ only to $(ij, ij \pm 1)$, and otherwise $i' = i$, $j' = j$. Since this compartmentalized matrix has the form $K_{ij,i'j'} = K(i-i', j-j')$, it can be diagonalized without difficulty. To do this, we introduce the unitary transformation

$$U_{ij\;rs} = \frac{1}{N} e^{\frac{2\pi i}{N}(ir+js)}, \text{ where } n = N^2,$$

for then

$$(U^+KU)_{rs,r's'} = \frac{1}{N^2} \sum_{ii'jj'} e^{\frac{2\pi i}{N}(i'r'+j's'-ir-js)} K(i-i', j-j')$$

$$= \frac{1}{N^2} \sum_{ii'jj'} e^{\frac{2\pi i}{N}(i'r'+j's'-(i+i')r-(j+j')s)} K(i,j)$$

$$= \delta_{rr'} \delta_{ss'} \sum_{i,j} K(i,j) e^{-\frac{2\pi i}{N}(ir+js)},$$

so that the diagonal 6 by 6 submatrices K^{rs} are obtained from the matrix displayed above by setting $\omega_1 = e^{\frac{2\pi i}{N} r}$, $\omega_2 = e^{\frac{2\pi i}{N} s}$. Let us take the limit as $n \to \infty$. If we define $\phi_1 = \frac{2\pi r}{N}$, $\phi_2 = \frac{2\pi s}{N}$ and $K^{rs} = k(\phi_1, \phi_2)$, then

$$\frac{1}{n} \log \det K = \frac{1}{N^2} \sum_{r,s} \log \det K^{rs} \underset{n \to \infty}{\longrightarrow}$$

$$\frac{1}{(2\pi)^2} \int_0^{2\pi} \int_0^{2\pi} \log \det k(\phi_1, \phi_2) d\phi_1 \, d\phi_2.$$

One readily finds that

$$\det k(\phi_1, \phi_2) = \frac{1}{v^4} [(1+v^2)^2 - 2v(1-v^2)(\cos \phi_1 + \cos \phi_2)]$$

where $v = \tanh \beta V$. Thus

$$-\beta f = \frac{1}{n} \log z = \log 2 - 2\beta V$$

$$+ \frac{1}{2} \frac{1}{(2\pi)^2} \int_0^{2\pi} \int_0^{2\pi} \log [\cosh^2 2\beta V - (\cos \phi_1 + \cos \phi_2) \sinh 2\beta V] d\phi_1 \, d\phi_2$$

which is Onsager's famous result.

Without carrying out an explicit evaluation, the analytic β dependence can easily be assessed. For high or low βV the argument of the logarithm is greater than zero and so f is analytic in βV. A break occurs when $\cosh^2 2\beta V/2 \sinh 2\beta V = 1$, for then the argument can vanish i.e. when $\phi_1 = \phi_2 = 0$. This ratio reaches its minimum value of unity at a unique value of βV, that given by $\sinh 2\beta V = 1$, or

$$e^{-2\beta V} = \sqrt{2} - 1.$$

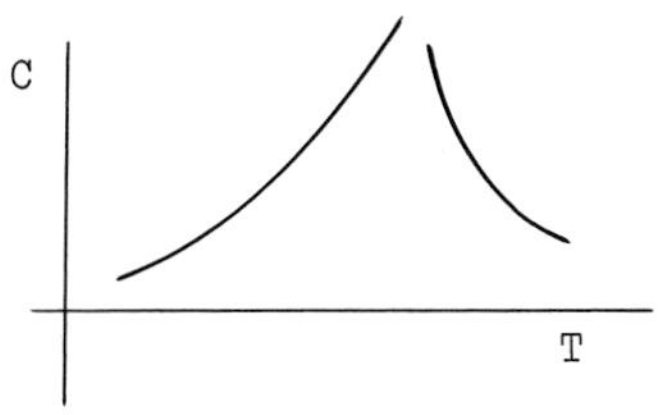

To find the singular behavior in the vicinity of this Curie point, we set $\delta = 1 - \sinh 2\beta V \sim -2(\cosh 2\beta V) \delta\beta V = -2^{3/2} \delta\beta V$. Then we have

$$\frac{1}{2}\frac{1}{(2\pi)^2}\iint \log\,[2-2\delta+\delta^2\ldots-(1-\delta)(\cos\phi_1+\cos\phi_2)]d\phi_1\,d\phi_2$$

$$=\frac{1}{2}\frac{1}{(2\pi)^2}\iint [\log\,(1-\delta)+\log\,(2-\cos\phi_1-\cos\phi_2+\delta^2+\ldots]d\phi_1\,d\phi_2$$

$$\sim\frac{1}{2}\frac{1}{(2\pi)^2}\iint \log\,(\frac{\phi_1^2+\phi_2^2}{2}+\delta^2)d\phi_1\,d\phi_2\sim\frac{1}{4\pi^2}\int\log\,(R+\delta^2)\,dR$$

$$\sim\frac{1}{2\pi}\delta^2\log|\delta|.$$

Hence the specific heat (in units of Boltzmann's constant) $C=-\beta^2\frac{d^2}{d\beta^2}\sim 8\beta^2V^2\frac{d^2}{d\delta^2}\;\frac{1}{2\pi}\delta^2\log|\delta|\sim\frac{8\beta^2V^2}{\pi}\log|\delta|$, which we have indicated in the diagram. Actually, one integral can be carried out, giving

$$-\beta f=\log\,(2\cosh\beta V)+\frac{1}{2\pi}\int_0^{\pi}\log\frac{1}{2}\{1+[1-(\frac{2\sinh 2\beta V}{\cosh^2 2\beta V})\sin^2\phi]^{\frac{1}{2}}\}\,d\phi,$$

the more usual elliptic integral form.

Finally, let us return to prove the Kasteleyn theorem which is crucial for the solution of all planar dimer packing problems. We first decompose the whole lattice into "polygons", each of which has no interior points, and such that any closed loop can be composed of these polygons. If the lattice is simply connected, it can be built up from these polygons in a suitable order such that each new polygon introduces at least one new bond. Consider the first polygon; its bond directions may obviously be chosen so that it is clockwise odd, i.e. a clockwise circuit of the polygon opposes an odd number of bonds. Since the second polygon introduces at least one new bond, the new bond signs can be chosen so that it too is clockwise odd. Continuing this process, we can assume that all of the elementary zero interior polygons are clockwise odd. Now consider a cycle on the lattice which is built up from elementary polygons. We shall show inductively that its clockwise parity is opposite to the parity of its interior, which will establish the theorem. Suppose that the clockwise parity of the (left) segment which has

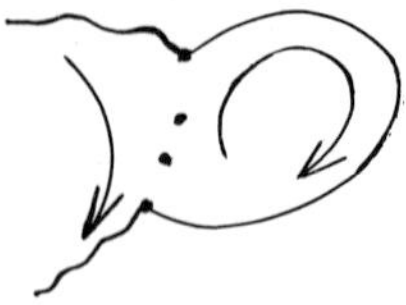

been built up is P and its interior I. Let us then join to it an elementary polygon, of parity one, adding $\mathcal{T}$ interior points in the process. Each of the $\mathcal{T}+1$ links which must be removed to carry out the combination must be clockwise with respect to one polygon, counterclockwise with respect to the other. Hence the combined clockwise parity is $P' = P + 1 - (\mathcal{T}+1) = P + \mathcal{T}$ while the combined interior $I' = I + \mathcal{T}$. Thus $P' - I' = P - I$. Since $P - I = 1$ for the first polygon, the induction is complete.

For a non-simply connected lattice, one chooses bond directions the same way, but consistency must then be verified in each case (e.g. a periodically bounded square lattice must have an even number of rows or columns).

4§. Other Combinatorial Solutions.

Other techniques yield a penultimate result which is simpler than a 6 by 6 matrix, but these appear intrinsically more complicated. Simplest is the original Kasteleyn method, applied again to the high temperature expansion, in which each bond is divided into three parts and the vertices then decorated as shown: For each of the four types of link configuration at a vertex, we then insert a dimer at each occupied arm and one to connect each pair

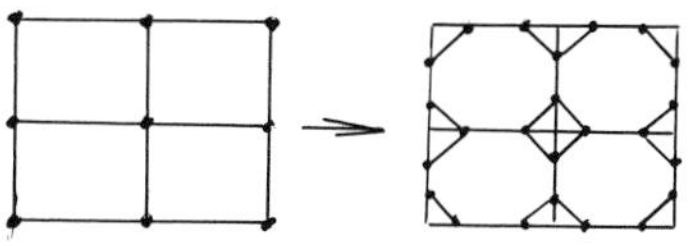

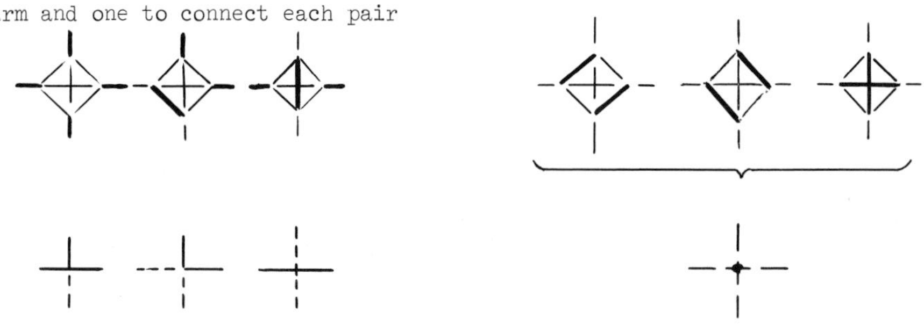

of unoccupied arms. There are two complementary drawbacks. First, there are three dimer configurations corresponding to an unlinked vertex, and second the lattice is not planar, so that Kasteleyn's theorem does not apply. However, if we adopt the bond directions shown, a loop of even interior which has no self-intersections does fit on a planar lattice and has odd clockwise parity, whereas each self-intersection contained in a loop is seen to reverse the clockwise parity. Thus, the

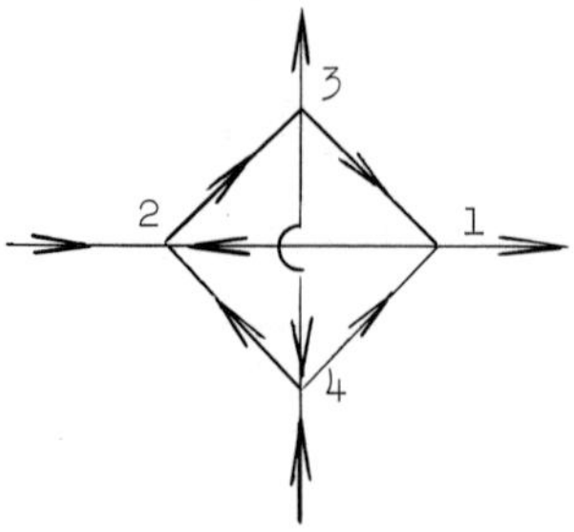

sign of each loop is given correctly, and in addition the weight of 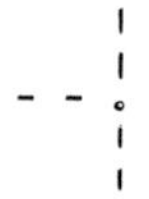 is $1 + 1 - 1 = 1$, as desired. In other words, we have at once for the square lattice

$$Z = 2^n e^{-2n\beta V} (\cosh \beta V)^{2n} \det K^{1/2}$$

where

$$K_{ij[i'j']} = \begin{pmatrix} 0 & 1+v\omega_2^{+1} & -1 & -1 \\ -1-v\omega_2^{-1} & 0 & 1 & -1 \\ 1 & -1 & 0 & 1+v\omega_1^{+1} \\ 1 & 1 & -1-v\omega_1^{-1} & 0 \end{pmatrix}$$

and $v = \tanh \beta V$. This of course gives the same result as Fisher's treatment.

In the approach of Kac and Ward (Phys. Rev. 88, 1332, 1952), the idea is to express the same series $\sum_r h(r)v^r$ as a permutation counting problem, hence as a permanent which is evaluated as a determinant. Thus the graphs have to be separated into closed disjoint cycles. Since two lines may cross at a vertex, we must choose as elements to be connected not the vertices, but the set of bonds. These can be enumerated, with orientation, by specifying the vertex

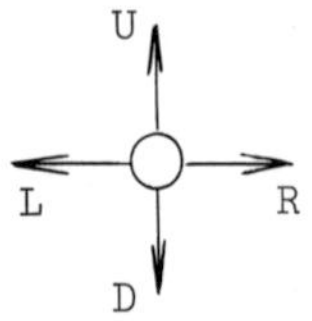

e.g. j and direction U, R, L, D. It will be convenient to allow for any weight v_{ij} per bond and hence, if $\ell(g)$ is the number of loops in g, to compute

$$\bar{Z} = \sum_{g \in G} (-1)^{\ell(g)} X(g),$$

where g is a member of the complete set of allowed graphs G and X(g) is the signed product of the weights of the bonds contained in g. There are two basic difficulties. First, separation of graphs into closed cycles is not unique, since at each crossover vertex, the compound loop can be separated in three different ways:

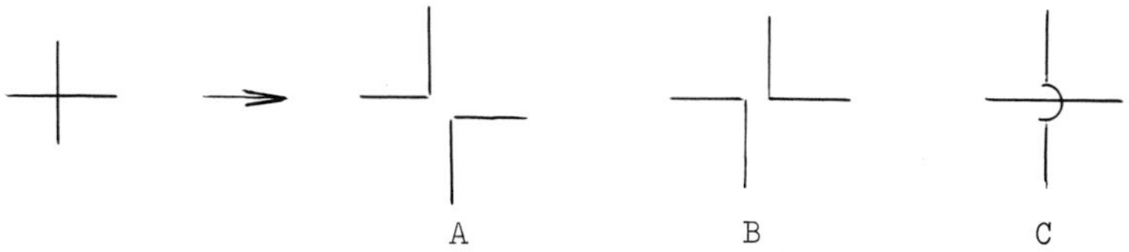

and any attempt at a unique separation (e.g. always use type C, or separate according to regions of + spin and regions of - spin) leads to severe problems. Second, the loops are not oriented, so that some process involving squaring the number of configurations will have to be used.

To overcome the first difficulty (Sherman, J. Math. Phys. 1, 202, 1960), we include all possible separations, with appropriate sign so that each desired graph ends up with an additional weight of 1. To see how this is done, consider all possible separations of the two simplest graphs with crossoverpoints:

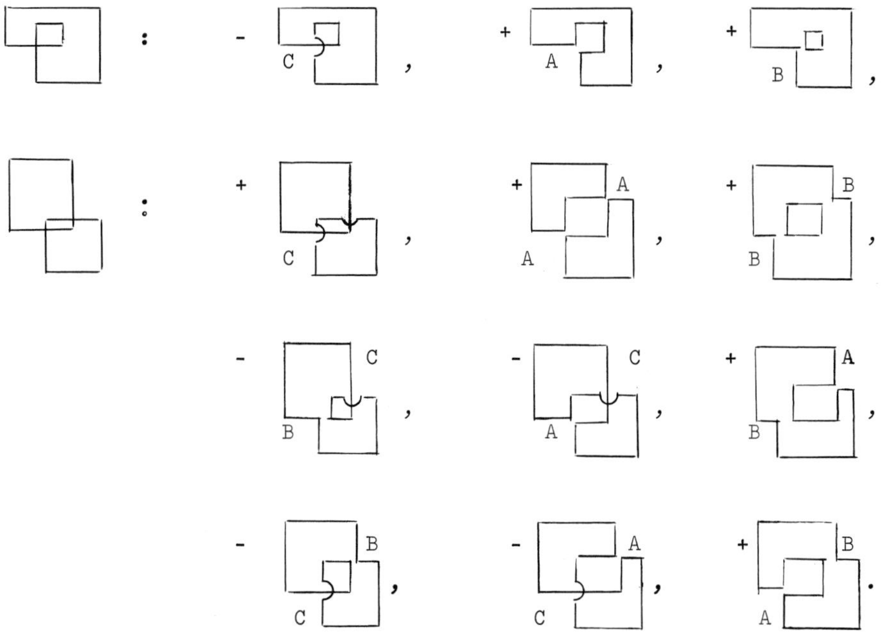

If, at each separation, type C is given a multiplicative weight of -1, type A and B of +1, the signs are as shown and each graph has a total weight of 1. This of course is completely general for to each separation point we are allocating a weight of 1 - 1 + 1 = 1. We see then that

$$\overline{Z} = \sum_{g \in G'} \gamma(g)\ (-1)^{\ell(g)} \chi(g)$$

where G' consists of all possible superpositions of disjoint bond loops and $\gamma(g) = \pm 1$ according to the parity of the number of type C separations. The remaining problem is to compute $\gamma(g)$ algebraically. For this purpose we observe that a vector which follows a simple loop turns by $\pm 2\pi$ in making a complete circuit. If a loop has one type C crossover, the circuit turns by $\pm 4\pi$ and in general each type C separation introduces a $\pm 2\pi$ into the total turning angle of all components of a graph. We conclude that

$$\overline{Z} = \sum_{g \in G'} e^{\frac{1}{2} it(g)} \chi(g)$$

where t(g) is the total turning angle of the graph.

A further generalization is required. We include loops of bonds which are not necessarily self-disjoint i.e. in which a bond can recur, as well as non-disjoint loops i.e. two loops with a common bond. These always cancel in pairs, as shown, one diagram possessing an additional perpendicular crossover, and hence do not contribute to the total. Thus

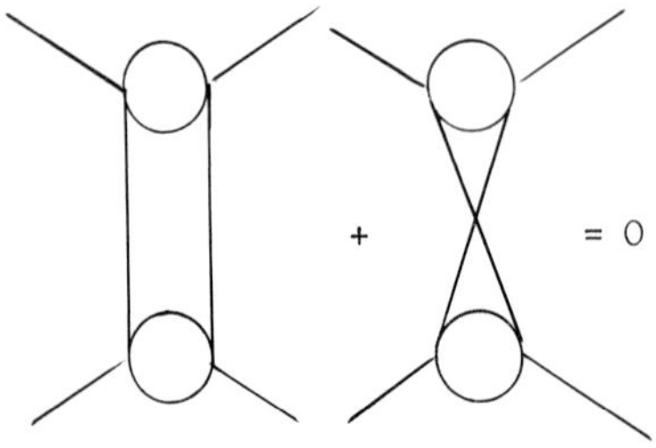

$$\overline{Z} = \sum_{g \in G''} e^{\frac{1}{2} it(g)} \chi(g),$$

where G'' includes all superpositions of oriented loops. By the argument associated with the above diagram, we can eliminate graphs in which any bond is described in the same direction more than once. Finally then

$$\overline{Z}^2 = \sum_{g \in G^*} e^{\frac{1}{2} it(g)} X(g)$$

where G^* consists of superpositions of non-disjoint oriented loops of directed bonds.

To represent $e^{\frac{1}{2} it(g)}$ explicitly, we need only associate $i^{1/2}$ to a clockwise 1/4 turn:

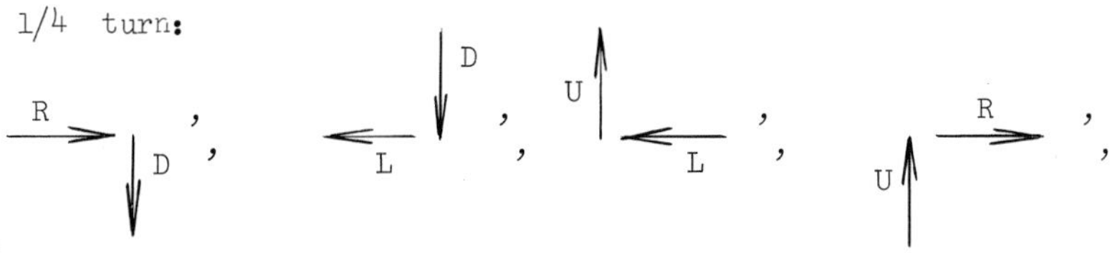

and $i^{-1/2}$ to a counter clockwise 1/4 turn. The transition matrix is then given by

$$K_{ij[i'j']} = \begin{pmatrix} \omega_1^{+1} v & 0 & i^{1/2}\omega_1^{+1} v & i^{-1/2}\omega_1^{+1} v \\ 0 & \omega_1^{-1} v & i^{-1/2}\omega_1^{-1} v & i^{1/2}\omega_1^{-1} v \\ i^{-1/2}\omega_2^{+1} v & i^{1/2}\omega_2^{+1} v & \omega_2^{+1} v & 0 \\ i^{1/2}\omega_2^{-1} v & i^{-1/2}\omega_2^{-1} v & 0 & \omega_2^{-1} v \end{pmatrix} \begin{matrix} R \\ L \\ U \\ D \end{matrix}$$

$$\qquad\qquad R \qquad\qquad L \qquad\qquad U \qquad\qquad D$$

Since our expression for $\overline{Z}^2$ contains within it the signed product of matrix elements (i.e. includes $(-1)^P$), det K will carry out the summation over all complete coverings of the bonds by cycles; however, we wish to include graphs which do not necessarily cover every bond. As is well known, det (I + K) is the sum of the determinants constructed from all subsets of indices. Hence we conclude that

$$\overline{Z}^2 = \det (I + K),$$

with of course the same result on evaluation as the previous two methods.

5§. Spin Correlations.

[Montroll, Potts, Ward J. Math. Phys. 4, 308, 1963; T. T. Wu, Phys.

Rev. 149, 380, 1966]

The finest, but unsystematized, data for the Ising model is the weight per configuration

$$\omega\{s_j\} = e^{-\beta E\{s_j\}} = e^{-\frac{\beta}{2}\Sigma' V_{pq}(1-s_p s_q)} e^{\beta \Sigma V_p s_p}.$$

The most gross information is the weighted counting

$$e^{-\beta F} = \sum_{\{s_j\}} \omega\{s_j\} \equiv Z,$$

sufficient for thermodynamics. Finer detail is available from the successive spin correlations, which are observable because they tell how the free energy F changes under a perturbation: if $E\{s_j\} \to E\{s_j\} + \delta E\{s_j\}$, then

$$\delta F = -\frac{1}{\beta}\, \delta \log \sum_{\{s_j\}} \omega\{s_j\} = -\frac{1}{\beta}\Big[\sum_{\{s_j\}} \delta\omega\{s_j\}/\sum_{\{s_j\}} \omega\{s_j\}\Big]$$

or

$$\delta F = \frac{\sum_{\{s_j\}} \delta E\{s_j\}\omega\{s_j\}}{\sum_{\{s_j\}} \omega\{s_j\}} = \langle \delta E\{s_j\}\rangle.$$

For perturbations affecting only one spin at a time, we require $\langle s_p\rangle = \sum_{\{s_j\}} s_p\, \omega\{s_j\}/\sum_{\{s_j\}} \omega\{s_j\}$ which is easily seen to vanish for the field free Ising model, and to be a constant when a constant external field is applied. Thus the effect of an incremental magnetic field: $\delta E = -\mu \sum_p B_p s_p$ is readily found. The first real structure appears in the pair correlation $\langle s_p s_q\rangle$, depending only upon the vector difference $p-q$ in the field free case, whose long range is responsible for spontaneous magnetization etc. The simplest way to obtain $\langle s_p s_q\rangle$ is to apply a controlled perturbation. We note that

$$\text{if } \delta E = \lambda\mathscr{E}, \text{ then } \langle \mathscr{E}\rangle = \frac{\partial F}{\partial\lambda}\Big|_{\lambda=0},$$

so that

$$\tfrac{1}{2}[1 - \langle s_p s_q \rangle] = \frac{\partial F}{\partial V_{pq}} ,$$

but this requires appending to the nearest neighbor Ising model a non-nearest neighbor potential, and is very difficult to carry out. Another approach is to apply a weak external field and look at its second order effect, that is if

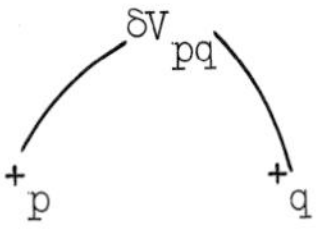

$$E = \frac{1}{2} \Sigma' V_{pq}(1 - s_p s_q) - \Sigma V_p s_p ,$$

then

$$Z\langle s_p s_q \rangle = \frac{1}{\beta^2} \frac{\partial^2 Z}{\partial V_p \partial V_q} ,$$

or

$$\langle s_p s_q \rangle = \frac{1}{\beta^2} \frac{\partial^2 \log Z}{\partial V_p \, \partial V_q} + \frac{1}{\beta^2} \frac{\partial \log Z}{\partial V_p} \frac{\partial \log Z}{\partial V_q}$$

$$= - \frac{1}{\beta} \frac{\partial^2 F}{\partial V_p \partial V_q} + \frac{\partial F}{\partial V_p} \frac{\partial F}{\partial V_q} ,$$

which requires only two external bonds (see Fisher).

The Ising model has one very special property which may be used to obtain the spin correlation from a perturbation which combines the advantages of only internal bonds and only nearest neighbor interaction. It is that since $s_j^2 = 1$ for all j, then if $p_0 = p$, $p_1, p_2, \ldots, p_{\alpha-1}$, $p_\alpha = q$ is any path on the lattice from p to q, we have

$$s_p s_q = \prod_{\gamma=1}^{\alpha} (s_{p_{\gamma-1}} s_{p_\gamma}).$$

Hence only nearest neighbor perturbations need be considered. This can be used in two distinct but of course ultimately equivalent ways. First, just as we used the identity

$$e^{\beta V s_p s_q} = (1 + s_p s_q \tanh \beta V) \cosh \beta V$$

to carry out a diagrammatic expansion of the partition function, we may also write

$$s_p s_q \, e^{\beta V s_p s_q} = (1 + s_p s_q \coth \beta V) \sinh \beta V.$$

Thus the only essential effect is a replacement of $v_\ell = \tanh \beta V_\ell$ by $1/v_\ell$ in the determinantal expansion based on bond connections. Second, we have the identity

$$\langle \prod_\gamma (s_{p_{\gamma-1}} s_{p_\gamma}) \rangle = \frac{1}{\beta Z} e^{-\frac{\beta}{2} \Sigma' V_{pq}} \frac{\partial^\alpha}{\prod_\gamma \partial V_{p_{\gamma-1} p_\gamma}} (Z e^{\frac{\beta}{2} \Sigma' V_{pq}}) \Big|_{\{V_{pq}=V\}}$$

Let us first examine nearest neighbor correlations, so that we only have to consider the bond $\ell = (p,q)$ for $|p-q| = 1$. We recall from Kasteleyn's approach that

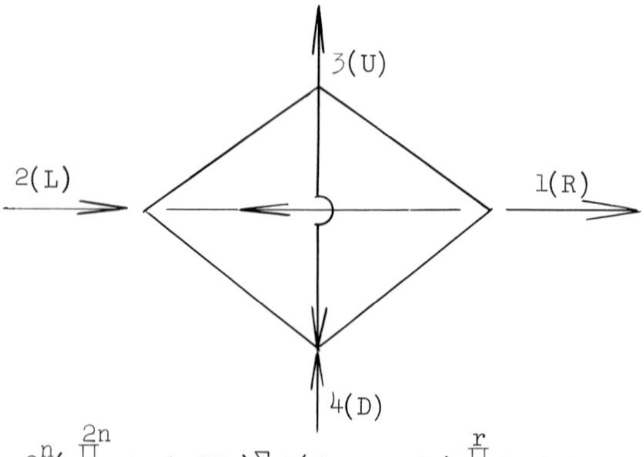

$$Z' \equiv Z e^{\beta \sum V_\ell} = 2^n \Big(\prod_{\ell=1}^{2n} \cosh \beta V_\ell \Big) \sum_r h(\ell_1, \ldots, \ell_r) \prod_{j=1}^{r} \tanh \beta V_{\ell_j}$$

$$= 2^n \Big(\prod_{\ell=1}^{2n} \cosh \beta V_\ell \Big) [\mathrm{Det}\ K]^{1/2},$$

where

$$K_{ij(i'j')} = \begin{pmatrix} 0 & 1+v_R\omega_2 & -1 & -1 \\ -1-v_L\omega_2^{-1} & 0 & 1 & -1 \\ 1 & -1 & 0 & 1+v_U\omega_1 \\ 1 & 1 & -1-v_D\omega_1^{-1} & 0 \end{pmatrix}$$

Clearly then

$$\langle s_p s_q \rangle = 1 - \frac{\partial F}{\partial V_{pq}} = - \frac{\partial F'}{\partial V_{pq}}\Big|_{V_{pq}=V}$$

$$= \frac{1}{\beta} \frac{\partial}{\partial V_{pq}} [\ \Sigma \log \cosh \beta V_\ell + \frac{1}{2} \log \operatorname{Det} K]$$

$$= v_{pq} + \frac{1}{2\beta} \frac{\partial}{\partial V_{pq}} \operatorname{Tr} (\log K)$$

$$= v_{pq} + \frac{1}{2\beta} \operatorname{Tr} (K^{-1} \frac{\partial K}{\partial V_{pq}}) = v_{pq} + \frac{1}{2} \frac{\partial v_{pq}}{\partial \beta V_{pq}} \operatorname{Tr}(K^{-1} \frac{\partial K}{\partial v_{pq}})$$

$$= v_{pq} + \frac{1}{2} (1 - v_{pq}^2) \operatorname{Tr} (K^{-1} \frac{\partial K}{\partial v_{pq}}).$$

To evaluate this expression we carry out the unitary transformation which diagonalizes K, thereby converting most of the matrix algebra to ordinary algebra. As we have previously seen, if

$U_{ij,rs} \equiv \frac{1}{N} e^{\frac{2\pi i}{N}(ir+js)}$ and K is periodic, translation invariant, then $(U^+KU)_{rs,r's'} = \delta_{rr'}\,\delta_{ss'}K[rs]$

where $K[rs] \equiv \sum_{\Delta_i,\Delta_j} K(\Delta_i,\Delta_j) e^{-\frac{2\pi i}{N}(r\Delta i+s\Delta j)}$.

Here of course we are using the notation $(p,q) = (ij,ij+1)$ and in the submatrix, $\omega_1 = e^{2\pi i\ r/N}$, $\omega_2 = {}^{2\pi i\ s/N}$. Hence

$$\operatorname{Tr}\left(K^{-1}\frac{\partial K}{\partial\beta V_{pq}}\right) = \operatorname{Tr}\,(U^{+}K^{-1}U)\left(U^{+}\frac{\partial K}{\partial\beta v_{pq}}U\right)$$

$$= \sum_{r,s} K^{-1}[rs]\left(U^{+}\frac{\partial K}{\partial\beta v_{pq}}U\right)_{rs,rs}$$

$$= \sum_{r,s} K^{-1}[rs]\, U^{+}_{rs,ij}\, \triangle_{ij,ij+1}\, U_{ij+1,rs}$$

$$+ \sum_{r,s} K^{-1}[rs]\, U^{+}_{rs,ij+1}\, \triangle_{ij+1,ij}\, U_{ij,rs}$$

$$= \sum_{rs} K^{-1}[rs]\frac{1}{N^2}\, e^{\frac{2\pi i}{N}s}\, \triangle_{ij,ij+1} + \sum_{r,s} K^{-1}[rs]\,\frac{1}{N^2}\, e^{-\frac{2\pi i}{N}s}\, \triangle_{ij+1,ij}$$

where

$$\triangle_{ij,ij+1} = \begin{pmatrix} 0 & 1 & 0 & 0 \\ 0 & 0 & 0 & 0 \\ 0 & 0 & 0 & 0 \\ 0 & 0 & 0 & 0 \end{pmatrix}, \quad \triangle_{ij+1,ij} = \begin{pmatrix} 0 & 0 & 0 & 0 \\ -1 & 0 & 0 & 0 \\ 0 & 0 & 0 & 0 \\ 0 & 0 & 0 & 0 \end{pmatrix}$$

Setting $\theta_1 = \frac{2\pi}{N}s$, $\theta_2 = \frac{2\pi}{N}r$ and taking the limit $N \to \infty$, we have

$$\frac{1}{(2\pi)^2}\iint [K^{-1}_{21}[\theta_1\ \theta_2]e^{i\theta_1} - K^{-1}_{21}[\theta_1\ \theta_2]e^{-i\theta_1}]d\theta_1\, d\theta_2$$

$$= \frac{2}{(2\pi)^2}\iint K^{-1}_{21}[\theta_1\ \theta_2]e^{i\theta_1}d\theta_1\, d\theta_2 .$$

But by a direct computation

$$K^{-1} = \frac{1}{(1+v^2)^2 - 2v(1-v^2)(\cos\theta_1 + \cos\theta_2)}$$

$$\begin{pmatrix} A-A^* & A+A^*-BAA^* & 2-BA^* & 2-BA \\ -A-A^*+B^*AA^* & A^*-A & -2+B^*A^* & 2-B^*A \\ -2+B^*A & 2-BA & B^*-B & B+B^*-BB^*A \\ -2+B^*A^* & -2+BA^* & -B-B^*+BB^*A^* & B-B^* \end{pmatrix}$$

where

$$A = 1 + ve^{i\theta_1}, \quad B = 1 + ve^{i\theta_2}, \quad v = v_{pq}.$$

Therefore

$$\langle s_p s_q \rangle = v + (1-v^2)\frac{1}{(2\pi)^2} \iint e^{i\theta_1} \frac{-(1-v^2)+ve^{-i\theta_2}(1+v^2+2v\cos\theta_1)}{(1+v^2)^2-2v(1-v^2)(\cos\theta_1+\cos\theta_2)}\, d\theta_1\, d\theta_2$$

$$= \frac{1}{(2\pi)^2} \iint \frac{2v(1+v^2)-v^2(1-v^2)e^{i\theta_2}-(1-v^2)e^{-i\theta_2}}{(1+v^2)^2-2v(1-v^2)(\cos\theta_1+\cos\theta_2)}\, d\theta_1\, d\theta_2$$

$$= \frac{1}{2\pi} \int v \frac{2\left(\frac{1+v^2}{1-v^2}\right) - ve^{i\theta_2} - \frac{1}{v}e^{-i\theta_2}}{\left(\left[\frac{1+v^2}{1-v^2} - 2v\cos\theta_2\right]^2+4v^2\right)^{1/2}}\, d\theta_2$$

$$= \begin{cases} (1+k)^{1/2}\left(\frac{1-k}{\pi} K(k) + \frac{1}{2}\right) & \text{for } k < 1 \quad (T < T_c) \\ (1+k)^{1/2}\left(\frac{1-k}{\pi k} K\left(\frac{1}{k}\right) + \frac{1}{2}\right) & \text{for } k > 1 \quad (T > T_c) \end{cases}$$

where $k \equiv \dfrac{1}{\sinh^2 2\beta V}$ and K is the complete elliptic integral.

In this special case of nearest neighbor correlations it is not necessary to examine the structure of the partition function in such detail. In fact it is clear that the nearest neighbor correlation is independent of which particular nearest neighbor we choose. Thus

$$\langle s_p s_q \rangle = - \left.\frac{\partial F'}{\partial V_{pq}}\right|_{V_{pq}=V} = -\frac{1}{2n} \sum_{p,q} \left.\frac{\partial F'}{\partial V_{pq}}\right|_{V_{pq}=V} = -\frac{1}{2n} \frac{\partial F'}{\partial V} \,,$$

which leads at once to the above result.

We now proceed to long range correlations, and our main purpose is to find the magnetization, which we recall is given by

$$I^2 = \lim_{|p-q| \to \infty} \langle s_p s_q \rangle .$$

We will restrict our attention to distant correlations along the same row, which we will compute by the $v \to \frac{1}{v}$ technique: if p and q are separated by m sites, then

$$\langle s_p s_q \rangle = v^m [\text{Det } K']^{1/2} [\text{Det } K]^{-1/2},$$

where the m bonds in question are given a weight of v^{-1} in K'. Let us write

$$K' = K + \Delta K,$$

where ΔK exists only for m compound rows and columns, on each of which it has the form (except at the corners)

$$\begin{pmatrix} 0 & (v-v^{-1})\omega_2 & 0 & 0 \\ (v^{-1}-v)\omega_2^{-1} & 0 & 0 & 0 \\ 0 & 0 & 0 & 0 \\ 0 & 0 & 0 & 0 \end{pmatrix}$$

Now

$$[\text{Det } K'][\text{Det } K]^{-1} = \text{Det } (K'K^{-1}) = \text{Det } (I + (\Delta K)K^{-1}).$$

Since $(\Delta K)K^{-1}$ has only m non-vanishing compound rows, and only one non-vanishing row in each compound of four, it is easy to see that $\text{Det } (I + (\Delta K)K^{-1})$ may be restricted to precisely these rows and columns. The evaluation of the elements of the determinant is readily carried out in the form

$$((\Delta K)K^{-1})_{ij,i'j'} = \sum_{r,s} ((\Delta K)U)_{ij,rs} \, K^{-1}[rs] U^{+}_{rs,i'j'},$$

which as $n \to \infty$ becomes a double integral once more. One of these integrals may be performed at once, as in the nearest neighbor case, and one finds after a certain amount of algebra that

$$\langle s_{11} s_{1m+1} \rangle = \begin{vmatrix} a_0 & a_1 & \cdots & a_{m-1} \\ a_{-1} & a_0 & & a_{m-2} \\ \vdots & & & \vdots \\ a_{-(m-1)} & & \cdots & a_0 \end{vmatrix}$$

where

$$a_r \equiv \frac{1}{2\pi} \int_{-\pi}^{\pi} e^{-ir\omega} e^{i\delta^*(\omega)} d\omega$$

and

$$e^{2i\delta^*(\omega)} \equiv \frac{(uv\, e^{i\omega}-1)(ve^{i\omega}-u)}{(e^{i\omega}-uv)(ue^{i\omega}-v)}$$

where

$$v = \tanh \beta V, \quad u = \frac{1-v}{1+v} = e^{-2\beta V}.$$

Note that when $m = 1$, $s_{11}s_{12} = a_0$ becomes identical with our previous result.

The asymptotic evaluation thus depends on the asymptotic properties of Toeplitz determinants. If the above determinant is designated as $D_m(a)$, it can be shown (Szegö's Theorem, see e.g. U. Grenander and G. Szegö, Toeplitz Forms and Their Applications, Univ. of Cal. Press (1958)) that

$$\lim_{n \to \infty} \frac{D_m(a)}{G^{m+1}} = e^{\sum_{n=1}^{\infty} nk_n k_{-n}},$$

where if

$$a_r = \frac{1}{2\pi}\int_{-\pi}^{\pi} e^{-ir\omega} {}_r i\delta^*(\omega)\, d\omega$$

then

$$G = \exp \frac{1}{2\pi}\int_{-\pi}^{\pi} \log f(\omega)d\omega$$

and

$$\sum_{n=-\infty}^{\infty} k_n e^{in\omega} = \log f(\omega).$$

In the present case $\log f(\omega) = i\delta^*(\omega)$ is an odd function so that $G = 1$. Furthermore $\log f(\omega)$ can readily be expanded out in powers of $e^{i\omega}$. One then finds that

$$I = \begin{cases} [1 = (\sinh 2\beta V)^{-4}]^{1/8} & \text{for } T < T_c \\ 0 & \text{for } T \geq T_c \end{cases}$$

which is Onsager's famous result.

INDEX

APPLIED MATHEMATICAL SCIENCES

Previously Published Volumes

Volume	ISBN	Title	Price
1	0-387-90021-7	Partial Differential Equations, F. John	$6.50
2	0-387-90022-5	Techniques of Asymptotic Analysis, L. Sirovich	$6.50
3	0-387-90023-3	Functional Differential Equations, J. Hale	$6.50
4	0-387-90027-6	Combinatorial Methods, J. K. Percus	$6.50

In Press

5	0-387-90028-4	Fluid Dynamics, K.O. Friedrichs and R. von Mises	$6.50
6	0-387-90029-2	A Course in Computational Probability and Statistics, W. Freiberger and U. Grenander	$6.50